7-8-95

To my Dear & Longtime friends

the Daltons,

Jim
Jody
Jack &
Jenna
+ sparky & George

Thanks for the years of support, fun, warmth, love & intelligence & more that we've shared, from the desert, + mountains to city & suburbs, between & beyond, no words can say enough. With love & Affection from your Pal

Michael S. Schneider

A Beginner's Guide to
Constructing the Universe

A Beginner's Guide to

Constructing

the Universe

The Mathematical Archetypes of Nature, Art, and Science

Michael S. Schneider

HarperCollins*Publishers*

Illustration credits appear on page 352.

HarperCollins books may be purchased for educational, business, or sales promotional use. For information, please write: Special Markets Department, HarperCollins Publishers, Inc., 10 East 53rd Street, New York, NY 10022.

FIRST EDITION

Designed by Nancy Singer

Library of Congress Cataloging-in-Publication Data

Schneider, Michael S., 1951–
    A beginner's guide to constructing the universe / by Michael S. Schneider. — 1st ed.
        p. cm.
    ISBN 0-06-016939-7
    1. Mathematics—Philosophy.   I. Title.
QA8.4.S374   1994
516'.001—dc20   92-56222

94 95 96 97 98 AC/RRD 10 9 8 7 6 5 4 3 2 1

To my parents,
Sally and Leonard, for their endless love,
guidance, and encouragement.

A universal beauty showed its face;
The invisible deep-fraught significances,
Here sheltered behind form's insensible screen,
Uncovered to him their deathless harmony
And the key to the wonder-book of common things.
In their uniting law stood up revealed
The multiple measures of the uplifting force,
The lines of the World-Geometer's technique,
The enchantments that uphold the cosmic web
And the magic underlying simple shapes.
—*Sri Aurobindo Ghose (1872–1950,*
Indian spiritual guide, poet)

Number is the within of all things.
—*Attributed to Pythagoras (c. 580–500 B.C.,*
Greek philosopher and mathematician)

The earth is rude, silent, incomprehensible at first,
nature is incomprehensible at first,
Be not discouraged, keep on,
there are divine things well envelop'd,
I swear to you there are divine beings
more beautiful than words can tell.
—*Walt Whitman (1819–1892, American poet)*

Education is the instruction of the intellect in the laws of
Nature, under which name I include not merely things and
their forces, but men and their ways; and the fashioning of the
affections and of the will into an earnest and loving desire to
move in harmony with those laws.
—*Thomas Henry Huxley*
(1825–1895, English biologist)

Contents

Acknowledgments

I'd like to express my sincere appreciation and thanks to the good people who produced this book at HarperCollins, especially my hardworking editors Rick Kot, Alice Rosengard, and Eamon Dolan, and copy editor Jeff Smith. I'd also like to thank Nancy Singer for helping celebrate geometry in the way she designed the book, and Gail Silver for her insightful page layout. And many thanks to my astute agents, Katinka Matson and John Brockman.

I especially wish to express my gratitude to John Michell, whose books, lectures, friendship, intelligence, humor, and vision have inspired me and many others to explore the harmony in numbers and shapes and who generously gave this book its main title and preface.

The works of many people from the deep past and immediate present have impassioned my interest in this subject, and I would like to acknowledge my debt to many including Keith Critchlow, John Anthony West, Robert Lawlor, Matila Ghyka, D. W. Thompson, T. A. Cook, Jay Hambridge, M. C. Escher, R. Buckminster Fuller, David Fideler, Jean and Katherine LeMay, Rupert Sheldrake, Rachel Fletcher, R. A. Schwaller de Lubicz, Sri Aurobindo, Pythagoras, and Plato, as well as the artisans and architects of ancient Egypt and Greece who knew how to make harmony visible.

For years of deeply interesting conversation, fun, friendship, love, and support I'd like to offer warm thanks and love to my brother, Jeffrey, and my parents and to my friends who live in the canyons on that island off the mainland, New York City, especially Naomi H. Cohen, Jim Pittman, Libby Reid, Barbara Crane, Charlie and Evelyn Herzer, Mark Hasselriis, and June Cobb of the Sacred Geometry Library at the Cathedral of St. John the Divine. To my friends embraced by the wide sky and natural order of Baker, Nevada, heartfelt thanks for the perpetual welcome and the tools with which to learn, including Ralph M. DeBit, Anita Hayden, and Jim Dalton. I'd also like to thank my friends, fellow educators, and students for sharing a dozen lovely years in the greenery of Gainesville, Florida.

I'd like to remember my teachers, who tried as best they could to get me to like math, and my many students in geometry workshops, from whom I have learned so much.

And I wish to express my wonder and thanks to nature, whose geometric jewelry adorns the world.

Geometry and the Quest for Reality

John Michell

Sooner or later there comes a time in life when you start thinking about Reality and where to find it. Some people tell you there is no such thing, that the world has nothing permanent in it, and, as far as you are concerned, consists merely of your fleeting experiences. Its framework, they say, is the random product of a natural process, meaningless and undirected.

Others believe that the world was made by a divine Creator, who continues to guide its development. This sounds a more interesting idea, but, as skeptics point out, every religion and church that upholds it does so by faith alone. If you are naturally faithful and can accept without question the orthodoxy of your particular religion or system of beliefs, you will feel no need to inquire further and this book will appear superfluous. It was written for those of us who lack or have lost the gift of simple faith, who need evidence for our beliefs. We cannot help being attracted by the religious view, that the world is a harmonious, divinely ordered creation in which, as Plato promised the uninitiated, "things are taken care of far better than you could possibly believe." Yet superficially it is a place of confusion and chaos, where suffering is constant and the ungodly flourish.

This is where we begin the quest for Reality. Looking closely at nature, the first insight we obtain is that, behind the apparently endless proliferation of natural objects, there is a far lesser number of apparently fixed types. We see, for example, that through every generation cats are cats and are programmed for catlike behavior. In the same way, every rose has the unique characteristics of a rose and every oak leaf is definitely an oak leaf. No two specimens of these are ever exactly the same, but each one is clearly a product of its formative type. If it were not so, if animals and plants simply inherited their progenitors' characteristics, the order of nature would soon dissolve into an infinite variety of creatures, undifferentiated by species and kinships.

This observation, of one type with innumerable products, gives rise to the old philosophical problem of the One and the Many. The problem is that, whereas the Many are visible and tangible and can be examined at leisure, the One is never seen or sensed, and its very existence is only inferred through the evident effect it has upon its products, the Many. Yet, paradoxically, the One is more truly real than the Many. In the visible world of nature all is flux. Everything is either being born or dying or moving between the two processes. Nothing ever achieves the goal of perfection or the state of equilibrium that would allow it to be described in essence. The phenomena of nature, said Plato, are always "becoming," never actually "are." Our five senses tell us that they are real, but the intellect judges differently, reasoning that the One, which is constant, creative, and ever the same, is more entitled to be called real than its ever-fluctuating products.

The search for Reality leads us inevitably toward the type, the enigmatic One that lies behind the obvious world of the Many. Immediately we encounter difficulties. Being imperceptible and existing only as abstractions, types cannot be apprehended by the methods of physical science. A number of modern scientists, perceiving the influence of types in nature, have attempted to bring them within the range of empirical study. Rupert Sheldrake, author of *A New Science of Life* and other works, has taken a bold step in that direction. In an earlier age, the Pythagoreans worked on systematizing the types by means of numerical formulae. Yet

Plato, who wrote at length on the subject of constant types (referred to as "forms"), was carefully ambiguous in defining them and never made clear the means by which they influence the world of appearances.

Plato did, however, give instructions on the procedure toward understanding the nature and function of the types. In the *Republic* he described the ascent of the mind through four different stages. It begins in Ignorance, when it does not even know that there is anything worth knowing. The next stage is Opinion, the stage in which TV chat-show participants are forever stuck. This is divided into two subcategories, Right Opinion and Wrong Opinion. Above that is the level of Reason. By education and study, particularly in certain mind-sharpening subjects, the candidate is prepared for entry into the fourth stage, which is called Intelligence (nous). One can be prepared for it but with no guarantee of success, for it is a level that one can only achieve on one's own, the level of heightened or true understanding, which is the mental level of an initiate.

The studies that Plato specified as most effective in preparing the mind for understanding are the so-called mathematical subjects, consisting of number itself, music, geometry, and astronomy. These were the main studies of Pythagoras and his followers, who anticipated the realization of modern physics in proclaiming that all scales and departments of nature were linked by the same code of number. Geometry is the purest visible expression of number. In Platonic terms, the effect of its study is to lead the mind upward from Opinion onto the level of Reason, where its premises are rooted. It then provides the bridge or ladder by which the mind can achieve its highest level in the realm of pure Intelligence.

Geometry is also the bridge between the One and the Many. When you draw one of its basic figures—a circle, say, or a triangle or regular polygon—you do not copy someone else's drawing; your model is the abstract ideal of a circle or triangle. It is the perfect form, the unchanging, unmanifest One. Below it are the Many—the expressions of that figure in design, art, and architecture. In nature also the One circle gives rise to the Many, in the shapes and orbits of the planets, in the roundness of berries, nests, eyeballs, and the cycles of time. On every scale, every natural pattern of growth or movement conforms inevitably to one or more of the simple geometric types. The pentagon, for example, lies behind each specimen of the five-petaled rose, the five-fingered starfish, and many other living forms, whereas the sixfold, hexagonal type, as seen in the structure of snowflakes and crystal generally, pertains typically to inanimate nature.

As soon as you enter upon the world of sacred, symbolic, or philosophical geometry—from your first, thoughtful construction of a circle with the circumference divided into its natural six parts—your mind is opened to new influences that stimulate and refine it. You begin to see, as never before, the wonderfully patterned beauty of Creation. You see true artistry, far above any human contrivance. This indeed is the very source of art. By contact with it your aesthetic senses are heightened and set upon the firm basis of truth. Beyond the obvious pleasure of contemplating the works of nature—the Many—is the delight that comes through the philosophical study of geometry, of moving toward the presence of the One.

Michael Schneider is an experienced teacher and, as you are entitled to expect, a master of his geometric craft. No one less qualified could set out its basic principles so clearly and simply. His much rarer asset is appreciation of the symbolic and cosmological symbolism inherent in geometry. That is the best reason for being interested in the subject, and it is the reason why the philosophers of ancient Greece, Egypt, and other civilizations made geometry and number the most important of their studies. The traditional science taught in their mystery schools is hardly known today. It is not available for study in any modern place of education, and there is very little writing on the subject. In this book you will find something that cannot be obtained elsewhere, a complete introduction to the geometric code of nature, written and illustrated by the most perceptive of its modern investigators.

Introduction

FANNING ANCIENT SPARKS

The universe may be a mystery, but it's not a secret. Each of us is capable of comprehending much more than we might realize. A vision of mathematics different from that which we were taught at school holds an accessible key to a nearby world of wonder and beauty.

In ancient Greece the advanced students of the philosopher Pythagoras who were engaged in deep studies of natural science and self-understanding were called *mathematekoi*, "those who studied all." The word *mathema* signified "learning in general" and was the root of the Old English *mathein*, "to be aware," and the Old German *munthen*, "to awaken." Today, the word *math* has, for most people, constricted its scope to emphasize mundane measurement and mere manipulation of quantities. We've unwittingly traded wide-ranging vision for narrow expertise.

It's a shame that children are exposed to numbers merely as quantities instead of qualities and characters with distinct personalities relating to each other in various patterns. If only they could see numbers and shapes as the ancients did, as symbols of principles available to teach us about the natural structure and processes of the universe and to give us perspective on human nature. Instead, "math education" for children demands rote memorization of procedures to get one "right answer" and pass innumerable "skill tests" to prove superficial mastery before moving on to the next isolated topic. Teachers call this the "drill and kill" method. Even its terminology informs us that this approach to math is full of problems. It's no wonder countless people are innumerate. We've lost sight of the spiritual qualities of number and shape by emphasizing brute quantity.

The most incomprehensible thing about the universe is that it is comprehensible.

—Albert Einstein (1879–1950)

Nothing in education is so astonishing as the amount of ignorance it accumulates in the form of inert facts.

—Henry Brooks Adams (1838–1918, American historian)

Bodily exercise, when compulsory, does no harm to the body; but knowledge which is acquired under compulsion obtains no hold on the mind.

—Plato (c. 428–348 B.C., Greek mathematical philosopher)

Wisdom should be cherished as a means of traveling from youth to old age, for it is more lasting than any other posession.

—Bias of Priene (c. 570 B.C., one of the Seven Sages of ancient Greece)

This book concerns mathematics, but not the kind you were shown at school.

The Roman goddess Numeria is said to have assisted in teaching each child to count. We must have had a misunderstanding because I grew up on uneasy terms with the subject of mathematics in school. I was intrigued by the infinity of numbers and could calculate mechanically if I had memorized the rule I was taught for a given situation. Although I liked science and art, math intimidated me. I remember my frustrated tears at age seven over the concept of subtracting by borrowing from another column when it contained a zero. I dreaded math, its mindless memorization and its tedious paperwork. (Pity the teachers who have to check it!) Math was dry and mechanical and had little relevance to my world. If we had looked at numbers to see how they behave with each other in wonderful patterns I might have liked math. Had I been shown how numbers and shapes relate to the world of nature I would have been thrilled. Instead, I was dulled by math anxiety and pop quizzes.

Fortunately for me, when I was sixteen one of my teachers mentioned that mathematics can be found in nature: a six-sided snowflake is shaped like a bee's cell and quartz crystal. I was stunned to make the connection. How could something as apparently irrelevant as mathematics be related to something as wonderful as nature? The ordinary world opened up to me and spoke the language of number and shape. No longer a foe, the dreaded mathematics became at once a teacher and a tool. Nature wasn't what it was made out to be, an antagonist to fear, conquer, and exploit, but a garden of wonder and a patient teacher worthy of great respect.

Over the years my studies led me to see that a profound understanding of number was prevalent in ancient times, more than is commonly acknowledged, and seamlessly wove mathematics, philosophy, art, religion, myth, nature, science, technology, and everyday life. This book is a fanning of some little sparks of philosophy from deep antiquity to introduce the general reader to another view of mathematics, nature, and ourselves that is our heritage and birthright. It requires that we liberate math from its pigeon-

Calvin and Hobbes

by Bill Watterson

hole and see it spanning the framework of the most inter-disciplinary topic possible: the universe.

Plato wrote that all knowledge is already deep within us, so no one can really teach us anything new. But we can remind each other of the archetypal principles of number and nature we already know but may have forgotten. This book is intended as that reminder to guide you through this world of wisdom, beauty, uplift, and delight and to remind you of the gentle, wise principles by which the universe is designed.

THREE LEVELS OF MATHEMATICS

Mathematics is whole, but it may be divided into three levels or approaches: secular, symbolic, and sacred.

Secular Mathematics

What is taught in school can be called secular mathematics. Adding the amounts of a store purchase, calculating change, weighing produce, measuring ingredients for a recipe, counting votes, telling time, designing a bookshelf or sky-scraper, measuring land boundaries, stock market econom-ics, calculating tax: we're taught these are the ends of math-ematics for nonscientists. We train children to be human pocket calculators. Even many educated people who are

By them [the Greeks] geometry was held in the very highest honor, and none were more illustrious than mathematicians. But we [Romans] have limited the practice of this art to its usefulness in measurement and calculation.

—Cicero (106–43 B.C., Roman orator, statesman, and philosopher)

adept at calculating have no idea that there is more to mathematics than mere reckoning of quantities.

This quantitative approach keeps us dull to the potential wisdom that the familiar counting numbers can teach us. When imaginatively taught to people beginning at an early age, mathematics can delight, inspire, and refine us. It can make us aware of the patterns with which the world and we are made. Instead, math is taught as a servant of commerce, without regard for its basis in nature. It is viewed as a distant subject that instills much more anxiety than wonder and inspiration. Mathematics is seen as outside us to be occasionally called upon, rather than woven into the fiber of our existence.

Symbolic Mathematics

Embedded in the very structure of ordinary numbers is another level, which may be called "symbolic" or "philosophical mathematics." It is well known that simple numbers and shapes relate to each other in harmonious recurring patterns. Just notice how floor tiles containing different shapes mesh. Mathematicians and scientists seek and study patterns as clues to a deeper understanding of the universe. What's more, symbolic mathematics recognizes numbers and shape patterns as representative of far-reaching principles. They can be guides to a deep cosmic canon of design. Nature itself rests on an internal foundation of archetypal principles symbolized by numbers, shapes, and their arithmetic and geometric relationships.

According to ancient mathematical philosophers, the simple counting numbers from one to ten and the shapes that represent them, such as circle, line, triangle, and square express a consistent, comprehensible language. The ten numbers are a complete archetypal sourcebook. They are the original ten patents for designs found all through the universe. These ideal patterns are the ones that were skewed and veiled in school and that nature approximates in all transitory forms, from the smallest subatomic particles to largest galactic clusters, crystals, plants, fruits and vegetables, weather patterns, and animal and human bodies. Any-

Nature, that universal and public manuscript.

—Sir Thomas Browne (1605–1682, English physician and author)

Nature is written in symbols and signs.

—John Greenleaf Whittier (1807–1892, American poet)

All things are full of signs, and it is a wise man who can learn about one thing from another.

—Plotinus (205–270, Roman Neoplatonic philosopher)

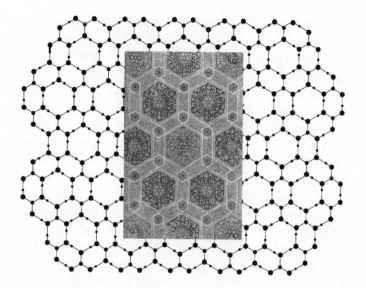

Islamic tiling pattern of a mosque and structure of the boric acid molecule show how identical shapes interlock in defined, recurring patterns in both art and nature.

I am not ambitious to appear a man of letters: I could be content the world should think I had scarce looked upon any other book than that of nature.

—Robert Boyle (1627–1691, British physicist and chemist)

In nature's infinite book of secrecy, a little I can read.

—William Shakespeare (1564–1616)

Living in the world without insight into the hidden laws of nature is like not knowing the language of the country in which one was born.

—Hazrat Inayat Khan (1882–1927, Sufi spiritual guide, musician, and writer

thing anyone can point to in nature is composed of small patterns and is a part of larger ones.

Historically, nature has been compared to a book written in geometric characters. In 1611 Galileo wrote:

> Philosophy is written in this grand book—I mean the universe—which stands continually open to our gaze, but it cannot be understood unless one first learns to comprehend the language and interpret the characters in which it is written. It is written in the language of mathematics, and its characters are triangles, circles, and other geometrical figures, without which it is humanly impossible to understand a single word of it; without these, one is wandering about in a dark labyrinth.

Reading the *Book of Nature* first requires familiarity with its alphabet of geometric glyphs. We're exposed to nature's text in the natural shapes around us, but we don't recognize it as something we can "read." Identifying shapes and patterns and knowing what principles they represent allows us to understand what nature is doing in any given situation and why these principles are applied in human affairs. Why are plates and pipes and planets round? Why do hurricanes and hair-curls, atoms and pinecones unfurl as spirals? Why

For the nature of number is to be informative, guiding and instructive for anybody in everything that is subject to doubt and that is unknown.
For nothing about things would be comprehensible to anybody, neither of things in themselves, nor of one in relation to the other, if number and its essence were nonexistent . . .
The essence of number, like harmony, does not allow misunderstanding, for this is strange to it.
Deception and envy are inherent to the unbounded, unknowable, and unreasonable . . . Truth, however, is inherent in the nature of number and inbred in it.

—Philolaus (Fifth century B.C., Greek Pythagorean philosopher)

didn't school teach us that when we see the same spiral shape in shells, galaxies, and watery whirlpools we are witnessing the principle of balance through motion? Or that the hexagonal cells of beehives package the maximum space using the least materials, energy, and time? Nature labels everything with a cosmic calligraphy, but we generally don't suspect even the existence of the language. It is an open secret, fully in view but usually unnoticed. Like consonants and vowels—like building blocks and growth patterns—numbers, shapes, and their patterns symbolize omnipresent principles, including wholeness, polarity, structure, balance, cycles, rhythm, and harmony. Each shape represents a different problem-solving strategy in the cosmic economy. To see more deeply into this design alphabet we must be conversant in nature's native tongue, the language of symbolic mathematics. This book is primarily concerned with symbolic mathematics.

Sacred Mathematics

The terms "sacred sites," "sacred geography," "sacred architecture," "sacred arithmetic," and "sacred geometry" seem overused today. To the ancients the "sacred" had a particular significance involving consciousness and the profound mystery of awareness. How are you and I aware of these very words and their significance? Now that you've read them, these words are no longer just on the page. They're within your awareness. Sacred space is within us. Not in our body or brain cells but in the volume of our consciousness. Wherever we go we bring the sacred within us to the sacred around us. We consecrate locations and studies by the presence of this awareness, not just the other way around. Why should the sites of stone temples and wonderful cathedrals be more sacred than a rocky desert or concrete city street if we bring holy consciousness to each of them? Geographic locations are no more or less sacred than any other, although they may be powerful telluric sites. Awareness is the ultimate sacred wonder. Why endow objects outside ourselves as sacred and ignore the same source within us? A surprising amount of the world's religious art and architecture has been designed using the timeless symbolic patterns of nature and number, but these patterns remain *symbolic of our own sacred inner realm,* symbolic of the subtle structure of awareness whose source is the same as archetypal number. All this was understood in ancient times and deemed so important that it was built into the culture on every level.

Modern science tells us that what we commonly call "reality" is a compilation of pictures based on a narrow sense-band view of surface features. The world we perceive is a small slice of a vast, mostly invisible energy-event. Mathematics can take us beyond our ordinary limits to the cosmic depths. Plato at his Academy required the study of mathematics as a prerequisite for *philosophia,* a term signifying "the love of wisdom" and "to lift the soul to truth." Just over a century earlier Pythagoras had invented the word "philosophy" as a result of a question posed to him. When asked "Are you wise?" he is said to have answered "No, but I'm a *lover* of wisdom." Both Pythagoras and Plato suggested that all citizens learn the properties of the first ten numbers as a form of moral instruction. The study and contemplation of number and geometry can show us, if we look with the eyes of ancient mathematical philosophers, that neither outer nature nor human nature is the hodgepodge it may seem. Symbolic mathematics provides a map of our own inner psychological and sacred spiritual structure. But studying number properties and intellectually knowing the road map, the symbolism, is not the same as actually taking the journey. We take that journey by finding within ourselves the universal principles these properties represent and by applying the knowledge to our own growth. We pay attention to paying attention, in imageless awareness, directing sustained attention to that which the symbols refer to within us. When the lessons of symbolic or philosophical mathematics seen in nature, which were designed into religious architecture or art, *are applied functionally* (not just intellectually) *to facilitate the growth and transformation of consciousness,* then mathematics may rightly be called "sacred." To me, the terms "sacred arithmetic" and "sacred geometry" only have significance when grounded in the experience of self-awareness. Religious art is sacred not only due to its subject matter but also because it was designed using the subtle symbolic language of number, shape, and proportion

"You amuse me," I said, "with your obvious fear that the public will disapprove if the subjects you prescribe don't seem useful. But it is in fact no easy matter, but very difficult for people to believe that there is a faculty in the mind of each of us which these studies purify and rekindle after it has been ruined and blinded by other pursuits, though it is more worth preserving than any eye since it is the only organ by which we perceive the truth. Those who agree with us about this will give your proposals unqualified approval, but those who are quite unaware of it will probably think you are talking nonsense, as they won't see what other benefit is to be expected from such studies."

—Plato

*The real mystery of life is
not a problem to be solved,
it is a reality to be
experienced.*

—J. J. Van der Leeuw

*The goal of life is living in
agreement with nature.*

—Zeno of Elea (490–430 B.C.,
Greek philosopher)

*We are born for
cooperation.*

—Marcus Aurelius (121–180,
philosopher at twelve,
Roman emperor at forty)

to teach self-understanding and functional self-development. Ancient Egyptian arts, crafts, and architecture perhaps provide the best accessible examples of design that used the symbolism of number, geometry, and nature to teach an accelerated form of self-development to trained initiates who knew how to translate the symbolism into meditative exercises.

Thus, true sacred geometry cannot be taught through books, but must remain as part of the ancient oral tradition passed from teacher to pupil, mouth to ear. Because over centuries this knowledge was passed along secretly so as not to conflict with the prevailing intolerant religious authorities or be disposed to those considered profane, there is still an aura of mysticism about it. But nature's patterns and those of our inner life are familiar to everyone and always available to us. The power which we seek is the power *with* which we seek. When we feel separate from the archetypes of nature, number, and shape we make them mystical, but this only keeps our selves in a mist. There is no need for secrecy and "occultism" any longer. These are everyone's life-facts. We can apply them to better appreciate the world, and we'll need them once we realize the urgency of cooperating with the way the world works. This book is concerned with dispelling the mysticism surrounding sacred mathematics by reintegrating the timeless knowledge of number and shape with its familiar expressions in nature, art, and its practical significance for self-development.

Studying, contemplating, and living in agreement with universal principles is a social responsibility and can be a spiritual path. It is becoming clear that when we cooperate with nature's ways we succeed; when we resist, we struggle. Implications for our environmental crises are obvious. Rather than an antagonist, nature can be our teacher to learn from and cooperate with to mutual benefit. To understand nature better, we first need to recognize the roles of its basic patterns.

MYTHMATICS: MATH AS CREATION MYTH

To say that ancient seers viewed the universe and themselves in a way different from ours is understatement. Today, we are emerging from the grip of literalism, mere quantitative measurement, and analysis. The ancients had a more poetic and synthesizing vision. We think we invented math, but the ancients knew that they were involved in a process of discovery. They personified in stories the forces of nature and gave them names of gods, goddesses, and nature spirits. Mythological exploits helped explain the origins and ways of familiar events from world creation to night and day, storms, agriculture, wars, love, hate, joy, sleep, life and death, and the full spectrum of natural and human affairs. But today we condescend and say that's how, in their childlike way, those pagans attempted to explain the mysteries of the world while lacking our "enlightened" scientific understanding. We think we really understand the forces of nature because we have labels, theories, terminology, experiments, measurements, advanced formulas, and astounding technological tools. With our habit of sensible literalism we have a difficult time believing that the ancient mythopoeic imagination could have been a valid tool for serious knowledge. Yet it is we who have the cruder understanding. By pigeonholing fragmented facts we miss the whole harmonious picture. We grope in a world we consider dangerous, accidental, and chaotic but one that is actually harmonious and awaiting our cooperation. If only we could see with the eyes of the ancients!

Scientists confirm with formulas what ancient seers knew through revelation: that the world's patterns and cycles are harmonious when seen as mathematical relationships. The ancients' intuitive revelations were grounded by the study of nature, numbers, and shapes. The ancient Greeks were particularly interested in the ways geometric shapes mesh in recurring harmonious patterns (the Greek word *harmonia* signifies "fitting together"), as we see in floor tiles, quilts, and wallpaper patterns, and how forces of nature orchestrate in small and large cycles (as satellites now reveal one worldwide weather pattern). Where we see "things," nouns and discrete objects, the ancient mathematical philosophers saw *processes,* verbs, transforming patterns meshed harmoniously. In truth, the "buck" never stops. The Hopi language retains this vision, having no nouns. (I'm not "wearing a shirt," but "I'm shirting.") Likewise, our word "cosmos" generally refers to "outer space." But the word derives from the Greek *kosmos* (signifying "embroidery"),

Divinity circumscribing the limits of the universe. Note his foot beyond the frame. (*Bible Moralisee,* c. 1250)

The Ancient of Days designing the universe through geometry. (Rockefeller Center)

which implied not a universe like a huge room filled with disconnected noun-things but the *orderliness* and *harmony* of woven patterns with which the universe is embroidered and moves. *Kosmos* signified the honorable and "right behavior" of the whole, the harmonious orchestration of the world's patterns and processes. In this original sense our word "cosmetics" refers not to the nouns involved, the lipstick and rouge, but to the process of bringing the elements of the face into harmony.

By studying the recurring harmonious patterns inherent in mathematics, music, and nature, ancient mathematical philosophers recognized that consistent correspondences occur throughout the universe. The Greeks investigated arithmetic by arranging pebbles in various geometric shapes ("figurate arithmetic") and so uncovered the archetypal patterns and principles inherent in many types of relationships. This study developed over centuries from what we call "number magic" to become the modern branch of mathematics called "number theory." Through extensive

Our biggest failure is our failure to see patterns.

—Marilyn Ferguson
(American writer)

Figurate Arithmetic. Simple arrangements of pebbles showed ancient mathematicians the patterns inherent in relationships among numbers. For example, any two consecutive "triangular" numbers (1, 3, 6, 10, 15 . . .) added together always make a "square" number (1, 4, 9, 16, 25 . . .).

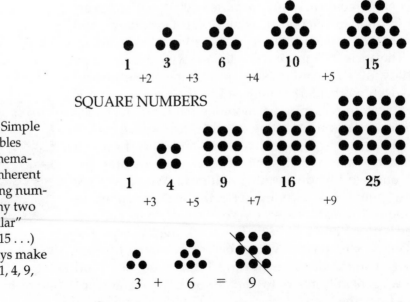

TRIANGULAR NUMBERS

1   3   6   10   15
+2   +3   +4   +5

SQUARE NUMBERS

1   4   9   16   25
+3   +5   +7   +9

3   +   6   =   9

study of patterns and by intuitive revelation the ancients realized that the structure and patterns of arithmetic and geometry reenact the creating processes found all through nature. In both number and nature they saw the same divine impress. Strip away all the sensory characteristics of color, texture, tone, taste, and smell from an object and only number remains as its size, weight, and quantity. Take away the features associated with number and it's all gone. Ordinary numbers and shapes represent eternal verities in a form made comprehensible to us. Mathematics is a philosophical language that reveals the Greek *archai,* the foundation principles of universal design and construction. Numerals, the written symbols of nontangible numbers, and geometric shapes, are the emblems of these archetypal principles. The archetypes are universal in that they are the same to everyone everywhere and in every era.

Nature's harmonious principles are exhibited through its mathematical relationships. As we learn to read its archetypal language, we discover that the topic of the book of nature is a story, a mythic quest about the process of transformation. This mathematical myth of the creative process, this "mythmatics," is not ancient or new-age but timeless, accessible in every age because it encodes eternal constants available to all. Although the myth's outer form adapts to different cultures, it may be rediscovered in any era by examining simple numbers, basic shapes, and ever-present nature. Thus, people of every historical period can understand the principles of nature's creating process by imaginatively examining the corresponding archetypal relationships inherent in mathematics, personified in mythology, and depicted in art and culture. This book is intended as a beginner's guide to the process of recovering that vision.

In the view of philosophical mathematicians, numbers and their associated shapes represent stages in the process of becoming. While integrated in the whole, each has a life of its own and a unique role in the cosmic myth. The archetypes of number and shape were personified in ancient cultures as various gods, goddesses, and world-builders. The classic myths elaborated on particular aspects of the universal principles they represented. Their interrelationships and liaisons, which we consider a "soap opera," transmitted

The harmony of the world is made manifest in Form and Number, and the heart and soul and all the poetry of Natural Philosophy are embodied in the concept of mathematical beauty.

—Sir D'Arcy Wentworth Thompson (1860–1948, Scottish zoologist, classical scholar)

The world, harmoniously confused,
Where order in variety we see,
And where, tho' all things differ,
All agree.

—Alexander Pope (1688–1744, English poet)

*A*ll *the effects of Nature are only the mathematical consequences of a small number of immutable laws.*

—Pierre Simon de Laplace
(1749–1827, French
astronomer and
mathematician)

*T*o *know all, it is necessary to know very little; but to know that very little, one must first know pretty much.*

—Georges I. Gurdjieff
(1872–1949)

*N*umbers *are the highest degree of knowledge. It is knowledge itself.*

—Plato

timeless truths. Through state sanction, as in Egypt, Greece, India, Africa, China, Tibet, Native America, and elsewhere, all aspects of society were organized according to a canon consistent with nature's own structure. The proportions of temples were always designed, situated, and built in accord with the number and shape symbolism representing the temple's deity. Through the process known as "gematria," which associates number values with letters of a carefully constructed alphabet, as in Greek, Hebrew, Arabic, Syriac, and Egyptian, the names, titles, and attributes of deities reveal, to those initiated to the code, their role in the cosmic constructive process. For example, seven was known as the number of the "virgin" because no number below seven enters into (divides) it, nor does it "reproduce" another number within the first ten. Through gematria, the Greek letters of the name of the maiden goddess Athena add up to seventy-seven, giving us a standard for measuring length, distance. The letters of her epithet, Pallas, add to 343 (which equals $7 \times 7 \times 7$), indicating volume. The letters of her appellation *parthenos* ("virgin") add to 515, and it is no coincidence that 51.5 degrees is extremely close to the angle within a regular heptagon, the shape with seven equal sides and angles, giving us area. Each "deity" represented a set of archetypal principles that were made comprehensible as mathematical patterns for practical use as length, area, volume, weight, duration, and music and were applied to the architecture and ritual of the deity's temple precinct. Each deity was responsible for overseeing part of the world's harmonious structure in the same way that his or her mathematical patterns did. Ancient mythology was consciously integrated with, and symbolized by, the universal and timeless canon of nature's mathematics.

This book makes practical the teachings of the Pythagoreans and others (including Plato, Theon of Smyrna, Iamblichus, Philolaus, Proclus, and Nichomachus) concerning the numbers one to ten as a complete frame for exploring and constructing the fundamental forms of the universe. These first ten numbers are like seeds from which all subsequent numbers and shapes grow, sharing and expanding their properties. The use of Greek words for these number principles, Monad, Dyad, Triad, and so on, as chapter titles, is intended for more than numerical order. These terms represent a process of unfolding cosmic qualities and principles. Each chapter is devoted to one number and shape principle. Although we count in sequence from one to ten, these number principles do not only unfold sequentially but interpenetrate the universe simultaneously in a cosmic symphony.

The cosmic creating process is deep within us and can emerge through our hands with the help of the three traditional tools of the geometer—the compass, the straightedge, and the pencil. To the ancients these tools represented three divine attributes that designed the patterns of the world. Just in this century modern science is rediscovering what the ancients symbolized by these three tools and their use. Albert Einstein referred to them as $E = mc^2$, the famous relationship among the universe's three "tools" of the configurating process: light, energy, and mass (or matter). The roles and motions of the geometer's three tools in geometric construction replicate the universe's own creating process whereby ideal patterns are approximated in natural design on all levels. We will look at our own constructions on paper as representing universal creations. Each step we take symbolizes a cosmic motion. By contemplating the essence of each of the ten numbers through our geometric constructions, we will gaze directly at the lines on the face of deep wisdom, the divine substrate of nature and ourselves.

ABOUT THIS BOOK

The best effect of any book is that it excites the reader to self-activity.
—*Thomas Carlyle (1795–1881, Scottish essayist and historian)*

Verifying number principles for ourselves is more important than just reading about them. No matter how conversant we are with the details of any map or restaurant menu, it's more informative to take the journey and more nourishing to eat the food. We can discover for ourselves the

Man, though seeing, suffered from blindness. . . And, for him, I found Numbers, the purest of inventions.

—Aeschylus (525–456 B.C., Greek tragic dramatist)

Geometry is knowledge of the eternally existent.

—Plato

He therefore who wishes to rejoice without doubt in regard to the truths underlying phenomena must know how to devote himself to experiment.

—Roger Bacon (1220–1292, English philosopher and scientist)

timelessness of nature's creating process by consciously participating in geometric activities.

By doing geometric constructions on paper, or with a stick in the sand, we are recalling the ageless process of creation, replicating with our mind and hands the generative principles by which the universe is evolving. Our constructions will demonstrate to us the harmonious principles at work in the world.

Symbolic and sacred mathematics encode subtle experiences whose purpose is different from that of secular mathematics. They can invigorate, refine, and elevate us. Our role as geometers is to discover the inherent proportion, balance, and harmony that exist in any situation. The study and experience of numeric and geometric proportion infuses in us an appreciation of proportion everywhere. The study of balance teaches us to recognize and seek a sense of balance in our lives. The study of harmony develops our sense of harmony in all relationships. Actually to see and work with unity and wholeness in geometry and natural forms, rather than just read about them, can help abolish our false notion of separateness from nature and from each other. It is this notion that ultimately fuels competition for the "goods of the earth" and contributes to environmental crises.

The material of this book developed over years of study and through holding public workshops and courses for people of widely varying interests, especially artists, architects, and educators. When my interest in this subject first began, I was sure that someone must have written one book explaining to my satisfaction the relationship between numbers, shapes, nature, science, art, and self. The information seems so fundamental, and people have been around for so many millennia to view it, that I was surprised to learn it only exists in books, articles, and fragments written mostly for specialized interests. I couldn't find the one book I was looking for that tied it all together, so I wrote it. It seems that the best approach to show you what I've learned is to take you beyond reading and involve you in the already-existing synthesis of geometry, natural science, art, and self-understanding through a "hands-on" approach. This book is intended to facilitate these studies in such a way as to see

If you do not rest upon the good foundation of nature, you will labor with little honor and less profit. [T]hose who take for their standard any one but nature—the mistress of all masters—weary themselves in vain.

—Leonardo da Vinci
(1452–1519)

ourselves as part of the overall harmony and have the tools to engage in it. It will guide you:

- To see how number and shape symbolism is already familiar to us through popular sayings, fairy tales, myths, and religious ritual.

- To examine the archetypal principles expressed by simple numbers and shapes and to let you verify the natural relationships existing among them. You'll do this by following step-by-step instructions for re-creating the basic geometric constructions symbolizing unfolding universal principles. You will use the tools of the geometer's creating process—the compass, straightedge, and pencil—to make two-dimensional and three-dimensional models of the principles expressed by the numbers from one to ten.

- To see where, and understand why, these numerical archetypes precipitate as organic and inorganic forms in nature and are the ones we rely upon in technology and engineering. You will be shown how to apply the geometric constructions to images of familiar natural forms discovering for yourself the ways nature speaks her geometric language.

- To examine examples of art throughout the world that display mathematical design. By applying the same geometric constructions to reproductions of art in the book (using tracing paper), you will learn to see how archetypal mathematical principles have been purposefully and consistently applied, to a surprising degree, to art, jewelry, and architecture throughout the world, endowing them with powerful compositional structure, symbolic significance, and intended psychological effects. Some examples will show the geometry the artist used. Further examples are left for you to complete and explore based on the geometric constructions you have learned.

NATURE HAS ALWAYS BEEN A MYSTERY TO ME

The real voyage of discovery consists not in seeking new lands but seeing with new eyes.

—Marcel Proust (1871–1922, French novelist)

*C*ome forth into the light
of things,
Let Nature be your teacher.

—William Wordsworth
(1770–1850, English poet)

This is not school, so doodling here at any time is okay, even encouraged. The word "doodle" signifies absent-minded scrawling and is related to "dawdle," wasting time in inconsequential activity. Both words derive from the Low German *Dudeldopp,* referring to a foolish or frivolous person. But Leonardo da Vinci and Dostoyevsky were doodlers. Doodling is a valid tool for awakening geometric vision and self-exploration. Try using your opposite hand to make you aware of your doodles in a different way. Psychologists call doodles manifestations of the unconscous. Letting our pencil flow without conscious intellectual control allows the archetypal patterns from within us to emerge. It is a geometer's method for self-discovery.

But don't just take my word for this. Try everything yourself. Keep as your own only what your experience validates.

As legal residents of the cosmos we have the authority to view the world's blueprints. Here they are. Construct the patterns that construct the universe.

CHAPTER
ONE

MONAD

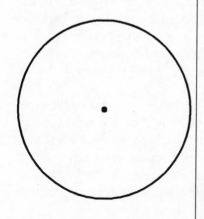

Wholly One

You cannot conceive the many without the one. . . . The study of the unit is among those that lead the mind on and turn it to the vision of reality.

—Plato

One principle must make the universe a single complex living creature, one from all.

—Plotinus

Everything an Indian does is in a circle, and that is because the power of the world always works in circles, and everything tries to be round.

—Black Elk (1863–1950)

The eye is the first circle, the horizon which it forms is the second: and throughout nature this primary figure is repeated without end.
    —Ralph Waldo Emerson (1803–1882, American essayist and poet)

It is a fallacy of the old schools to divide man into parcels, elements, thoughts, emotions, intuitions, etc. All human faculties consist of an interconnected whole.
        —Alfred Korzybski (1879–1950, American semanticist)

All are but parts of one stupendous whole.

—Alexander Pope

Whence shall he have grief, how shall he be deluded who sees everywhere the Oneness?

—Isha Upanishad

THE CIRCLE DRAWS US

In the fourteenth century Pope Benedictus XII was selecting artists to work for the Vatican, requesting from each applicant a sample of his ability. Although the Florentine painter Giotto (1266–1337) was known as a master of design and composition, he submitted only a circle drawn freehand, the famous "O of Giotto." Yet he was awarded the commission. Why? What's so impressive about a simple circle?

Give a young child a crayon and paper and observe what he draws. At the earliest ages children scrawl lines and zigzags. There comes a time when they discover that a line's end can meet its beginning, and they take delight in the loop. It continues endlessly around and creates an inside separate from an outside. Eventually, they come upon the circle. The circle brings the loop to perfection, so round in every direction. Children love to trace circular objects like cups and cans to achieve the fascinating perfection so hard for a young hand to draw.

The discovery and appreciation of the circle is our early glimpse into the wholeness, unity, and divine order of the universe. Some psychologists say that the discovery of the circle arrives as the child discovers the self and distinguishes himself from another. Even as adults our attention remains hypnotically drawn to circles, toward their centers, in objects we create and those we see. We draw circles and they draw us.

Looking at a circle is like looking into a mirror. We create and respond irresistibly to circles, cylinders, and spheres because we recognize ourselves in them. The message of the shape bypasses our conscious mental circuitry and speaks directly to the quiet intelligence of our deepest being. The circle is a reflection of the world's—and our own—deep perfection, unity, design excellence, wholeness, and divine nature.

To ancient mathematical philosophers, the circle symbolized the number one. They knew it as the source of all subsequent shapes, the womb in which all geometric patterns develop. The Greek term for the principles represented by the circle was Monad, from the root *menein*, "to be stable," and *monas*, or "Oneness." In the alphanumeric correspon-

Ancient Hindu priests used circular mandalas for practice in concentrating the mind. Although they have quite a different purpose in mind, modern graphic designers of commercial logos also know that circles irresistibly lure our attention toward their center.

dences of the Greeks, the letters of the word *monas* add up to 361. The system allows a difference of one, so the word for "oneness" becomes 360, not coincidentally, as that is also the number of degrees around a full circle. Ancient mathematical philosophers referred to the Monad as The First, The Seed, The Essence, The Builder, The Foundation, The Space-Producer and, most dramatic, The Immutable Truth and Destiny. It was also called Atlas because like the Titan upholding the heavens it supports, it connects and separates the numbers it produces.

The ancient philosophers conceived that the Monad breathes in the void and creates all subsequent numbers ($111111111 \times 111111111 = 12345678987654321$). Numbers only express different qualities of the Monad. The ancients didn't consider unity to be a "number" but rather a parent of numbers. They noted that unity exists in all things yet remains inapparent. They saw the relation of the Monad to all numbers in a metaphor of simple arithmetic: any number when multiplied by unity remains itself ($3 \times 1 = 3$). The same is true when unity divides into any number ($\frac{5}{1} = 5$). Unity

The *ouraboros* (Greek for "tail-biter"), a circular snake with its tail in its mouth, appears as a symbol on every continent representing eternity, not as all time but time-*lessness*.

The Hindu deity Shiva Nataraj, arguably the world's oldest continuously worshipped deity, represents the transforming life force that motivates all forms. Shiva is shown here dancing the universe into circular manifestation.

always preserves the identity of all it encounters. We might say that "one" waits quietly within each form without stirring, motionless, never mingling yet supporting all. The Monad is the universe's common denominator. The ancient Gnostics called it the "silence force." The universe was carved of this primeval silence. Everything strives in one way or another toward unity.

There's more to a circle than just a curved line. It's a wonderful first glyph of nature's alphabet. Every circle is identical. They only differ in size. Each circle you see or create is a profound statement about the transcendental nature of the uni-verse. Expanding from the "nowhere" of its dimensionless center to the infinitely many points of its circumference, a circle implies the mysterious generation from nothing to everything. Its radius and circumference are never both measurable at the same time in similar units due to their mutual relation to the transcendental value known as "pi" = 3.1415926 . . . When either the radius or circumference is measurable in whole, rational units, the other is an endless, irrational decimal. Thus, a circle represents the limited and unlimited in one body.

Our deepest awareness, the power that motivates all

In my end is my beginning.

—Motto of Mary, Queen of Scots (1542–1587)

The Tibetan Wheel of Life resembles a Christian image of God holding the cosmos. (Engraving after the fresco by Piero di Puccio, c. 1400)

The Creation of the World and the Expulsion from Paradise depicts the Garden, or "divine state," at the center of circular rings. (Fifteenth-century Italian painter Giovanni di Paolo)

It is Unity that doth enchant me.

—Giordano Bruno (1548–1600, Italian philosopher)

Beatrice and Dante gaze into the concentric rings of light nearest God.

awareness, which we can call the "Power to Be Conscious," of which we are not ordinarily cognizant, recognizes its own transcendental nature in the geometry of the circle. For this reason the circle has been a universal symbol of an ideal perfection and divine state that always exists around and within us whether we acknowledge it or not. Religious art has traditionally turned to the circle to symbolize this state of divinity as "heaven," "paradise," "eternity," and "enlightenment."

Giotto's perfectly drawn circle communicated this universal ideal.

IMPORTANT TIPS ON USING THE GEOMETER'S TOOLS

Before we use the three classic tools of the geometer, here are some points you should consider:

1. These three tools—compass, straightedge, and pencil—are ancient, found in various forms in most

Line upon line, line upon line; here a little and there a little.

—Bible (Isaiah 28:10)

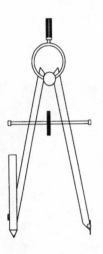

cultures. Used by artists, architects, and crafts-people, they are both practical and symbolic.

2. Whether you use a metal compass or a string tied to a stick in the earth, these tools represent divine attributes. Treat them with respect.

3. Do nothing unconsciously. Be aware of each action you perform with them. No act in a geometric construction is trivial or without profound symbolism and correspondence to the world's creating process.

4. Don't erase mistakes. Just as we cannot undo life's missteps, leave all marks where you make them and live with them until you can do the construction differently.

Compass

The Medieval geometers contemplated the compass as an abstract symbol of the eye of God. In their worldview its legs represented rays of light and grace shining from heaven to earth, from deep within us outward toward the periphery of our ordinary awareness. The compass has only one role: from a central seed-point it creates the transcendental hole called the circle. It opens up a divine space of light, awareness, and potential configuration. Remember that every cir-

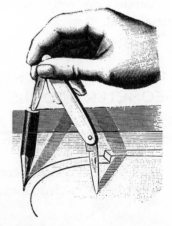

cle you construct represents the Monad, the complete universe. When turning a circle, hold the compass at its top, not its leg, to avoid changing the size of the circle. Turn each circle in one complete rotation without hesitation, leaning the compass forward as you turn. Avoid drawing only arcs and partial circles in the constructions. Complete all circles to see the full pattern of forces.

Straightedge

Where circles intersect, points are created. A straightedge allows us to rule lines between points, revealing paths of energy, motion, and force between them. Your straightedge needs no measure-marks. Measure is built into the performance of the geometry.

The edge should be free of nicks. When drawing a line, put your pencil on one point and bring the straightedge against it, pivoting it to the second point. Then put the pencil on that second point to "test" its accuracy against the edge. Hold the straightedge firmly. Draw the line toward yourself in one steady stroke. Imagine the emerging line as a path of energy.

Pencil

The scribe system, whether pencil and paper or stick and earth, is the means by which an archetypal pattern materializes. Pencil and paper translate divine, eternal ideas into symbols accessible to the geometer's sight. When you commit to creating a pencil stroke, move firmly, steadily, and deliberately without hesitation.

Keep the point sharp.

Making a Point

Creating one central point from which to radiate a circle is the beginning of all geometric construction, no matter how complex it becomes. The steps involved in any construction are a metaphor for the stages of the divine, ongoing creating process itself. Anywhere you place a center you can scribe a circle and symbolically create the space of the universe itself.

The circle is a shape begging to be organized. As the value "one," or unit, is the parent of all numbers, the circle

*T*rue Thou art, lad, as the circumference to its center.

—Herman Melville
(1819–1891, American novelist)

The circle accommodates all shapes within itself.

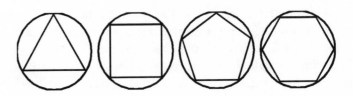

The world is single and it came into being from the center outwards.

—Joannes Stobaeus (fifth century A.D., Greek anthologist)

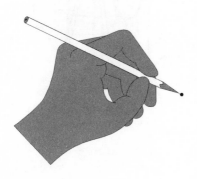

God makes himself known to the world; He fills up the whole circle of the universe, but makes his particular abode in the center, which is the soul of the just.

—Lucian (c. 240–312), Christian theologian)

is the parent of all shapes. All subsequent shapes and patterns will be inscribed within this all-encompassing Monad. That's why the world is called a *universe* (Latin for "one turn"). No other universes exist except within this One, which Plato refers to as the "whole of wholes."

To construct a symbolic universe, begin with a point, the circle's essence. Hold the closed compass upright, and poke a symbolic point upon the paper. A true point is impossible to draw, having absolutely no dimension, not length, width, or height. Ours is not a true mathematical point because seen under the microscope it's a three-dimensional heap of carbon. Ideal geometry is impossible to perform with human tools. We can only symbolize ideals by the geometric shapes we construct. The closed upright compass on the point represents the mythological world axis, world mountain, or holy center of many cultures, the symbolic pole or spine that supports creation and around which everything turns in adoration. Traditionally, the center is the most honorable place, known to the Greeks as "the keep of Zeus." Protector of hearths and boundaries (centers and circumferences) and the source of moral order, Zeus dispensed judgment from the center.

Nothing exists without a center around which it revolves, whether the nucleus of an atom, the heart of our body, hearth of the home, capital of a nation, sun in the solar system, or black hole at the core of a galaxy. When the center does not hold, the entire affair collapses. An idea or conversation is considered "pointless" not because it leads nowhere but because it has no *center* holding it together.

The point is the source of our whole of wholes. It is beyond understanding, unknowable, silently self-enfolded. But like a seed, a point will expand to fulfill itself as a circle.

Geometric construction can be used as a form of meditation. Ponder the point as a seed enfolding a sacred mystery. Hold that sense of wonder through the full construction.

We can use this approach to symbolic geometry as sacred geometry by *experiencing* what the construction process symbolizes within us. We can do this by shifting attention from outside to inside, from the page to its viewer.

Consider the point as symbolizing your own "center." What center? Dancers and gymnasts gracefully work with the body's center of gravity to balance during motion. Undersea divers walk around at the bottom wearing a twenty-pound weight at their groin, the center of gravity. Even when sitting motionless everyone has a psychological center of gravity, the thoughts, emotions, or desires with which we identify and from which we view the world at any given moment. Through meditation and self-contemplation we can seek a more subtle center, our higher or deeper self, *the power that motivates* the actions, emotions, thoughts, and desires. This center of gravity is not in space but in pure awareness, the "place" in you now aware of these very words. Where do you "hear" these words? You may believe or doubt that there is such a center, but you cannot doubt the *power* with which you doubt. This power is the motivating power-with-which-we-are-conscious, identical to the heart of every natural form and symbolized by the center of the circle. It is the seed of our mysterious ability to be aware. Only during psychological stillness, with no mental, emotional, or desire ripples disturbing the quiet "pond" of awareness within us, can we consciously approach our own deep sacred center. When you mark a point on paper, contemplate it as representing a center of profound awareness about to expand, construct, and motivate the universe. Then look inward, silence the voices, and seek that center's correspondence within you.

TO THE MARGIN DANCE

The point is made, the center established. The compass stands upright on it. Now we open the compass. This seemingly trivial act is an important stage in the geometric

*T*he one Godhead, secret in all beings, all-pervading, the inner Self of all, presiding over all action, witness, conscious knower and absolute . . . the One in control over the many who are passive to Nature, fashions one seed in many ways.

—Swetaswatara Upanishad

*A*s on the smooth expanse of crystal lakes, the sinking stone at first a circle makes; The trembling surface by the motion stirr'd, Spread in a second circle, then a third; Wide, and more wide, the floating rings advance, Fill all the wat'ry plain, and to the margin dance.

—Alexander Pope

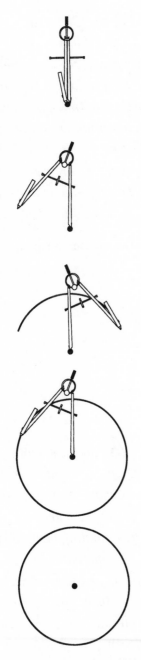

metaphor of the cosmic creating process. It represents the first archetypal principle of the Monad: equal expansion in all directions.

In many myths, the universal creating process begins with an expansion from a divine center, like the very first Biblical command, "Let there be Light." In Hindu mythology, the dimensionless Brahma speaks aloud the word *aham*, "I Am," a word made of the first, middle, and final letters of the Sanskrit alphabet, which represents the circle's three parts; the center, the radius, and the circumference, and our own spiritual center, psychological reaches, and outer material form. The opening compass represents the first manifestation of God's light and Brahma's voice, illuminating

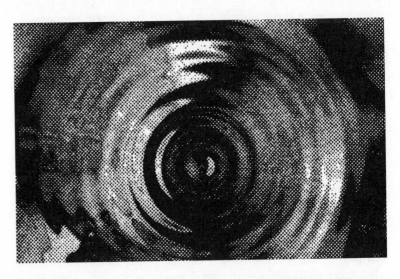

A point within a circle was the Egyptian, Chinese, and Mayan glyph for "light."

Ripples expanding in a pond present an optical illusion. We assume that the water is moving outward but it's not. What we see is a wave of *energy* from the impacting pebble racing outward through the water equally in all directions. Watch a floating object as waves pass by. The object only bobs up and down while the wave traverses horizontally. Visible matter merely makes the energy pattern visible. In a small puddle, attraction among the water molecules holds the wave together. Large ocean waves are held together by gravity.

and vibrating the universe into existence, as expanding states of self-awareness, which we call "nature."

Nature's forms represent invisible forces made visible. The force of the circle's equal expansion works through different materials. Tap the side of a round cup of liquid, and watch as perfect concentric rings appear and converge to the center, then pass it and expand outward again. Nature delights in the principle of equal expansion in concentric ripples, splashes, craters, bubbles, flowers, and exploding stars. As you open your compass, consider that you are metaphorically repeating this first principle of the Monad, the opening of light, space, time, and power in all directions.

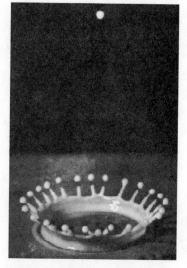

Impact ("splash") crater of milk, like a crater on the moon, is a diagram of expanding forces. The energy spreads as ripples on a pond but is limited by the material. The splash of every sphere-striving raindrop speaks the principle of the Monad.

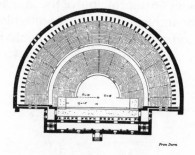

A Greek amphitheater's circular design matches the shape of the invisible but expanding sound waves of the actors' voices and provides everyone with an equal view.

The wave is not the water. The water merely told us about the wave moving by.

—R. Buckminster Fuller (1895–1983, American inventor, engineer, poet, mathematician, and futurist)

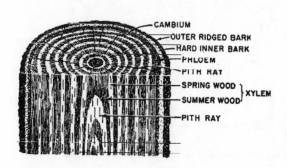

CAMBIUM
OUTER RIDGED BARK
HARD INNER BARK
PHLOEM
PITH RAY
SPRING WOOD
SUMMER WOOD } XYLEM
PITH RAY

Tree stump. Since plants are composed mostly of water (organized by minerals), it's natural that their growth rings resemble watery ripples expanding very slowly across a pond.

THE WHEEL WHIRL'D

Behold the world, how it is whirled round,
   And for it is so whirl'd is named so.
                    —*Sir John Davies (1569–1626, English jurist and poet)*

The second principle of the Monad is expressed by the circle's rotary motion. Unlike the still center, the circumference speaks of movement. We replicate this universal principle in our geometric constructions whenever we turn the compass around its point and scribe a circle. Symbolized in nearly every culture as a wheel, the circle represents nature's universal cycles, circulations, circuits, orbits, periodicities, vibrations, and rhythms.

Because cycles are a principle of the Monad, they are all-pervasive in the universe. We are thoroughly enmeshed in cycles and periodic rhythms but notice only the most obvious, like our breath and hunger or the time or season.

Cycles characterize biological and inorganic processes in all organisms from simplest to most complex. The life cycles of plants and animals, from seed to fruit, all migrations, metamorphoses, hibernations and reproductive cycles are synchronized with seasonal periodicities.

12-2. *Cycle of development of the bean plant, from seed to seed.*

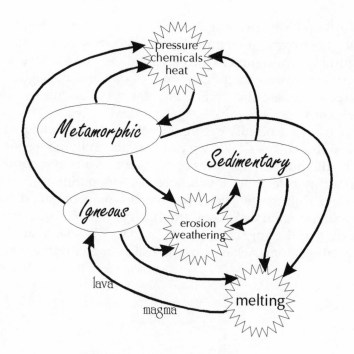

The Rock Cycle. Consider the slowly changing rocks that mesh their own long cycles of weathering, melting, cooling, crushing, compression, and reformation within a greater whole. The global circulation of weather, as well as the planetary water, oxygen, and nitrogen cycles are also well known.

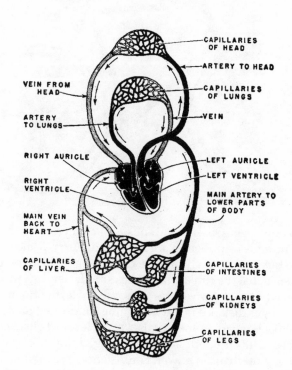

Circulatory System. Our heart is at the "center" of concentric rings.

*E*ven the seasons form a
great circle in their
changing, and always
come back again to where
they were. The life of a
man is in a circle from
childhood to childhood and
so it is in everything where
power moves. Our teepees
were round like the nests of
birds, and these were
always set in a circle, the
nation's hoop.

—Black Elk

All cycles have rising and declining phases. When one side goes up, the other goes down. This is true on any scale, in turning wheels and in the rising/falling pulse of an emotional burst, the changing amount of daylight through the seasons, the rise and decline of great cultures, and the life cycles of stars.

Observe a rapidly cycling bicycle wheel or ceiling fan. When it revolves slowly we can see each individual spoke or blade. But when it turns faster our nervous system just cannot register the revolutions as discontinuous, and beyond about 1,550 cycles per second the spokes and fan appear as a solid disk. Look around at any "solid" object. Be aware that the appearance of its "surface" is due to rapidly oscillating atoms, which move so fast as to give our nervous

Illusion of wheels. Observe the spokes of a bicycle wheel as it spins. The ends near the rim are rapid and blurry while the ends near the hub move slowly and are clearly seen. The paradox of the wheel is that the rim actually does spin faster than its hub because, while they turn across the same angle, the rim must travel a longer distance in the same time. To do so it moves faster. Wheels, gears, cranks, dials, knobs, levers, belts, and ball bearings use this property for magnifying, diminishing, and transferring mechanical power. For instance, a revolving door is easier to push toward its outside than near the center pivot.

system the *impression* of a smooth surface. The same is true for our sense of hearing. A card held against turning bicycle spokes or the teeth of a turning gear will produce discrete sounds until the rapidity is such that the sound is perceived as a continuous hum. All senses are fooled by rapid cyclic vibration so that even texture, smell, and taste appear as continuous to our registration faculties.

Every process is characterized by cycles. The appearances of the entire world with all its natural and technological cycles are images rooted in the archetypal cyclic principle of the Monad, represented by the geometer's turning compass. Cooperating with nature requires that we recognize the existence, and learn the ways, of its omnipresent cycles.

*N*o! no arresting the vast
wheel of Time,
That round and round still
turns with onward might.

—Charles C. Clarke
(1787–1877, English critic and
scholar)

*M*y career had begun in
Washington and it would
end there. I liked the idea of
a circle being completed.

—Helen Hayes (1900–1993,
American actress)

Astronomical cycles. The nested cycles of the moons, planets, suns, and galaxies are documented by astronomers. Moons orbit planets which revolve around suns. Spinning star clusters group as revolving galaxies, which whirl among other spinning galactic clusters, parts of cycles beyond comprehension. Here, as seen through time, the stars seem to trace concentric circles around the Pole Star.

*T*he sage is he who has
attained the central point
of the wheel and proves it
without himself
participating in the
movement and remains
bound to the Unvarying
Mean.

—Taoist saying

Cyclic Time. Instruments of time reckoning from sundials and calendars to atomic clocks are structural symbols of astronomical, atomic, and biological cycles. Hands of the analog (round) clock model the apparent path of the sun through the sky, or the Earth's daily spin seen from above the South Pole.

Round Aztec calendars (and those of other cultures) were arranged on the ground as models of their cyclic calendar, socially and symbolically meshing the structure of time with that of space in rounds of agricultural rituals, fairs, and festivals around a circle of locations through the year.

Asar (in Greek and English Osiris), the Egyptian personification of cycles. In general, gods and goddesses of the ancient world were personified archetypal principles whose actions manifest themselves in nature. Asar, often depicted green-faced (symbolizing vegetation), personifies all cycles in nature. Here he revolves in the "waters" of chaos, lifting the sky goddess Nut into spiritual light.

PERFECT SPACE FOR A UNIVERSE

A circle may be small yet it may be as mathematically beautiful and perfect as a large one.
—*Isaac D'Israeli (1766–1848, English man of letters)*

The wheel has come full circle.

—William Shakespeare

The third all-pervasive principle of the Monad involves the area *within* the circumference. A circle is not just the curve but the miraculous space inside, which manifests

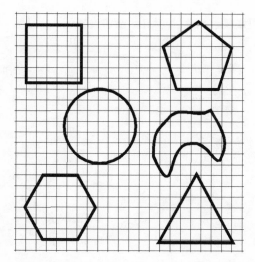

A simple experiment with a single loop of string and graph paper, chessboard, or checkered tablecloth will prove the supremacy of a circle. Form the loop into different shapes and lay each over the squares of the graph paper. Count the number of squares each shape encloses. Although each shape has the identical perimeter because the loop's length doesn't change, only the circle covers the most squares.

between nothingness (zero-dimensional point) and everything (infinitely many points around the circumference).

A circle expresses the most practical and efficient geometric space for natural and human creations to occur. Of all shapes, a circle encloses the most space by the smallest perimeter. In other words, the most enclosure with least exposure. The Monad's third principle is maximized efficiency.

Look for circles, cylinders, and spheres and you'll see them everywhere:

Shields. A round shield gave the ancient soldier maximum protection behind the largest area while employing the least material and having the least weight.

Manhole covers. Did you ever wonder why a manhole cover is round? A circle is the only shape that won't fall into its own hole. A square or triangular cover could drop in along its diagonal, which is longer than any side, but a circle's diameter is everywhere the same.

Ring roads. Ring roads or beltways around every major city and town allow a traveler easiest access into the center of a city

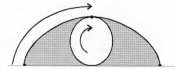

The moving path of a fixed point on a rolling circle leaves an arched trail. Maintaining unity, the two shaded portions on each side of the circle have the exact area as the rolling circle that produced them. Verify this by taping a pencil to a metal can and rolling it along the bottom of a wall at a right angle to graph paper on the floor.

When a wagon train formed a circle for protection it enclosed the most ground while offering the least exposure to attack from outside.

The 'Grands Projets' of Paris

Great Arch of La Défense

La Villette (A park, a museum of science and industry and a music center)

AVE. JEAN JAURES

AVE. DES CHAMPS-ELYSEES

BLVD. HAUSSMANN

BOIS DE BOULOGNE

Grand Louvre

Orsay Museum

Bastille Opera

Arab World Institute

Ministry of Finance

BLVD. PERIPHERIQUE

Seine

Bibliothèque de France

Ring road around a city.

Circles are prais'd not that abound In largeness, but th' exactly round: So life we praise, that doth excel Not in much time, but acting well.

—Edmund Waller (1606–1687, English poet)

from anywhere on the periphery. A city, a cell, and a nation are considered independent when they control their own borders.

Cups and Cans. A cylindrical cup or can contains more material within it than a square- or hexagonal-shaped container, made of the same amount of material, does. The circular shape allows it to hold more using the least material and weight. Only a sphere would be more efficient (but impractical to stack).

Plates. A round dinner plate holds more food than other shapes while allowing the least opportunity for it to fall over the side.

Pizza. A round pizza uses the least amount of dough to fill the largest space and provide the greatest area for toppings.

Round tables.   A round table is an expression of the equality of the diners. Its maximized surface area allows them to share more dishes than any other shape with the same perimeter would.

Bird's nest/Igloo/African hut. Animal and human builders in all climates prefer circular and hemispherical homes that enclose the largest living space with the minimum exposure while using the least materials, time, and energy to build.

*Consider then how lofty
and how wide
is the excellence of the
Eternal Worth
which in so many mirrors
can divide
its power and majesty
forevermore,
Itself remaining as One, as
It was before.*

—Dante (1265–1321)

*Exile is terrible to those
who have, as it were, a
circumscribed habitation;
but not to those who look
upon the whole globe as
one city.*

—Cicero

The Monad, or oneness, expressed as a point and a circle, is the foundation for our geometric construction of the universe. The three parts of the circle—center, circumference, and radius forming the space within—correspond to the three principles of the Monad: equal expansion, cycles, and efficient space. These principles, along with the Monad's wholeness, are all-pervasive and lie at the foundation of the world's objects and events, as the number one is hidden within every integer.

The Monad is knowable to us through its expression in nature's designs and human affairs as equal expansion, cycles, and efficient space. Natural structures are universally recognized as beautiful and most efficient. We, too, are part of the world's harmonious design and can't help but express the Monad's principles in the things we do and create.

Everything seeks unity. The goal of many religions and mythic ordeals is to return to a lost state of Divine Oneness. But we have no need to *return* to a state of oneness because unity is axiomatic and we *already* are integrated in it. Barely recognizing our situation, here and now we live in a whole and beautifully harmonious wonder world. Only a self-imposed illusion of separateness keeps us from recognizing our own center of awareness and identity with the One. To understand this unity the ancient mathematical philosophers contemplated the principles of the Monad through the arithmetic principles of the "number" one and by exploring its geometric expression as the circle.

This illusion of separateness is not a direct characteristic of the Monad but requires twoness. The Two, an "other," proceeds from the One for the process of universal construction. To understand how duality and the sense of separation derive from unity and work in the world and ourselves, we must go deeper into the mythic realm of geometry and number to the principle of twoness, known to the Greeks as the Dyad.

Chapter Two

Dyad

It Takes Two to Tango

The opposite is beneficial; from things that differ comes the fairest attunement; all things are born through strife.
—*Heraclitus (c. 540–c. 480 B.C., Greek philosopher)*

In the Two we experience the very essence of number more intensely than in other numbers, that essence being to bind many together into one, to equate plurality and unity. Our mind divides the world into heaven and earth, day and night, light and darkness, right and left, man and woman, I and you—and the more strongly we sense the separation between these poles, whatever they may be, the more powerfully do we also sense their unity.
—*Karl Menninger (1893–1990, American psychiatrist)*

This whole wide world is only he and she.
—*Sri Aurobindo Ghose*

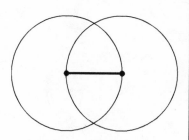

Birth of the Line.

THE BIRTH OF THE OTHER

The circle is the womb and cradle of our symbolic universe. It shows us the characteristics of unity, everywhere the same and containing no differences within it. Like a pebble tossed into a pond, a circle can only reproduce more circles in its own likeness. The ancient mathematical philosophers saw this in the metaphor of arithmetic. They noticed that no matter how many times unity is multiplied by itself, the result is the same: one ($1 \times 1 \times 1 \times \ldots \times 1 \times 1 = 1$). So how *does* unity, oneness, step beyond itself and become the many? How can the Monad generate the other principles, other shapes, other numbers? How does the "same" produce an "other" How does the primeval "I" generate its "Thou"?

With a mirror. It simply needs another circle identical to itself. The circle replicates a mate for itself by contemplating itself, reflecting its light, and casting its own shadow.

The process is mirrored through geometry as the birth of the line. It has been done from time immemorial through an ancient geometric construction called in Western tradition the *vesica piscis* (Latin for "bladder of the fish"). Try this: (1) First mark a center and spin a circle with the compass. (2) Lift the compass *without changing the size of its opening* and reverse it. Place its sharp point anywhere on the circumference to make a new center and swing another circle. This

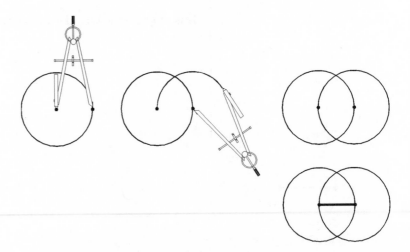

A line is born within the *vesica piscis*.

second center is actually an emanation of the original circle since it resides on its circumference. (3) These intersecting circles, linked across their centers to form a line, make an ancient and obvious symbol of twoness. The overlapping space between them is the *vesica piscis*.

As if across a mirror, the archetypal line is drawn between the two centers.

Sit up straight and get a sense of your spine as a line. To a symbolic geometer a line is a picture of energy, tension, force, action, impulse, urge, direction, movement. The straight lines we will draw in our constructions represent the tension and motion between the poles of every creating process.

A "true" archetypal line is one-dimensional, having length but no height or thickness. It simply expresses the principle of extension. Like a "true" point a "true" line is impossible to draw with material tools. But these two, the point/circle and the line, are the parents of all subsequent geometric constructions that *do* manifest themselves in actual nature.

Phantom circle. With your face a comfortable distance from this circle, relax your eyes and let your focus drift until the circle seems to separate into two side-by-side circles. When they seem to touch each other's center, adjust and hold the focus. This illusion is the result of one awareness viewing through two separate eyes.

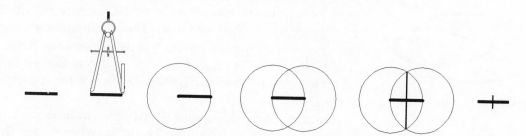

Using the *vesica piscis* to divide a line in half.

THE DUAL THRONG

The principle of "twoness," or "otherness," was called Dyad by the Greek philosophers of the five centuries before Christ. They were suspicious of it because it seemed to revolt from unity, distancing itself from the divine Monad. They referred to the Dyad as "audacity" for its boldness in implying a separation from the original wholeness and "anguish" due to its

Rumor doth double, like the voice and echo, the numbers of the feared.

—William Shakespeare

Couples are wholes and not wholes, what agrees disagrees, the concordant is discordant. From all things one and from one all things.

—Heraclitus

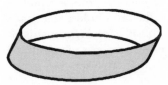

Loop: two separate sides.

Möbius strip: one-sided.

Möbius strip: two as one. Give a strip of paper a half twist and tape the ends together. This Möbius strip has only one side and one edge, although it appears to be two-sided. Trace your finger around the face and you'll return to the same beginning. It's a three-dimensional symbol of the Dyad, of one appearing as two.

inevitable yearning to return to unity. It was also called "distress," "falling short," "the lie," and "illusion" since they believed the Monad alone was all. Today, we employ this negative aspect in the derogatory phrases "two-faced" and "speaking with a forked tongue."

The principle of the Dyad is polarity. Polar tension occurs in all natural and human affairs as any opposing relationship, contrast, difference. It is at the root of our pernicious notion of separateness from each other, from nature, and from our own inherent divinity.

The paradox of the Dyad is that while it appears to separate from unity, its opposite poles remember their source and attract each other in an attempt to merge and return to that state of unity. The Dyad simultaneously divides and unites, repels and attracts, separates from unity and craves to return to it. A line creates both a boundary that divides and a link that binds. We know we are under the sway of the Dyad when we are attracted or repelled by anything.

The linguistic roots of the word "two" support its dual nature. *Separation* is emphasized in words having the prefix "di-" or "du-": discord, difference, dispute, dissent, disunion, difficult, dilemma, divide, distinct, distance, distinguish, disgruntled, distribute, diverge, dichotomy, doubt, duo, deuce, duet, dual and duel. Through linguistic evolution this root also gave us the words of personal polarity "thou" and "you." Our sad and dangerous notion of separation from nature and each other is rooted in the principles of the Dyad.

On the other hand, another twoness indicator, "tw-," implies *joining*, as in two, twin, twain, between, twine (two twisted threads), and twilight (a blend of day and night). There seem to be more words emphasizing "separation" than "joining," implying an alienation from unity.

Polar tension is at the root of all birth and creation. As the popular phrase reminds us, it takes two to tango. Exactly two people of opposite gender, no more or less, can produce a child. When cool, dry air penetrates warm, wet air, rain precipitates. Woven cloth manifests itself at the intersection of warp and woof. Two poles of a battery, positive and negative, are needed to complete an electric circuit. Two fixed ends of a guitar string allow us to pluck it, creating vibration, sound, and music. One chopstick is motionless, the other moves, and together they can pick up food. There isn't anything composed of matter (or antimatter) that avoids polarity. Even the geometric compass operates by the interplay of two legs, one motionless and the other moving, the poles of center and circumference.

The Dyad is the basis of every creative process. It shows up as rhythmic oscillation between opposite poles, as close as our beating heart and as far as quasars pulsing at the edge of the universe.

Human nature mirrors outer nature. All personal relationships have at their essence the archetypal tension between opposites. Polarities drive the dances of mating, love and hate, taking responsibility or assigning blame, strength and tenderness; they are integral to political opposition parties, diplomacy, business partnerships and business rivalries. Psychologists tell us that within each of us are the poles of anima and animus, female and male aspects whose relative balance determines how we relate to other men and women. Although we think we act independently, we follow nature's polar principles in most everything we do.

A look at language reveals our investment in polarization. How often do we hear the words good/evil, true/false, hot/cold, always/never, win/lose, pass/fail, input/output, us/them, this/that, right/wrong, high/low, physics/metaphysics, man/nature, contract/expand, constructive/destructive, visible/invisible, over/under, body/mind, thesis/antithesis, up/down, laugh/cry, permanence/change, virtue/vice, presence/absence, pressure/release, joy/despair, health/disease, too little/too much, success/failure, creature/Creator, pleasure/pain, profit/loss, limited/unlimited. Our private world is filled with these twin labels and the pictures and values that accompany them. This is the great "dual throng" of language. Every entity confronts itself with its own opposite. For each Tweedledum there must be a Tweedledee. Neither word in any pair can exist alone. Each implies the other. If there's something you don't like, you can assume that its opposite exists, which

Everybody's shouting "Which side are you on?"

—Bob Dylan (1941–)

There is no quality in this world that is not what it is merely by contrast. Nothing exists in itself.

—Herman Melville

Everything that originated from the tree of knowledge carries in it duality.

—*Zohar* (mystical Jewish text)

Every language having a structure, by the very nature of language, reflects in its own structure that of the world as assumed by those who evolved the language. In other words, we read unconsciously into the world the structure of the language we use.

—Alfred Korzybski

you will like. But shunning one while chasing its opposite only invokes and strengthens the one we try to avoid. Accepting one gives equal strength to its opposite. The words we speak, write, and think will unconsciously help determine what we see; thus, our language can shepherd us into defending ideas we may not actually hold. But experience doesn't support two-valued orientation. Life-facts are in between. That's why the ancients called Dyad "illusion."

Under the sway of the Dyad we see walls, boundaries, dividers. Polarized thinking encourages our sense of separateness and deflects our vision from the world's—and our own—inherent unity. We cannot live in a sense of oneness, unity, wholeness, completeness while bouncing in the mirror world of implied opposites, continually attracted or repelled, feeling separate from each other, from nature and from a deep relation to our inner selves. This notion of separateness from nature, in fact, encourages the environmental crises we find ourselves in. We would not attempt to tame, master, plunder, and foul nature if we recognized how those acts affect ourselves. To avoid disaster we must learn to think differently and avoid the traps of polarity in everyday language. Even the most seemingly innocuous statement blinds us to what's really going on. For example, when we say, "This world is not real; reality is somewhere else," or "Heaven is my home, not this world," or "Ideal geometric shapes exist in an archetypal realm, beyond our knowing," we're ensuring that we won't find what we seek. By thinking that what we seek lies elsewhere, we miss seeing it here right before us. Ideals aren't elsewhere but are embedded in the event. We don't notice because we're lured to one pole or the other. Only by acknowledging both poles in the pair as inseparable can we overcome relative duality and get to their common source in the Monad. In our deepest Self we are beyond all polarity.

IT'S A CHECKERBOARD WORLD

Many worldwide religions and mythic cosmologies include stories of the birth of a primeval dichotomy, usually lovers, enemies, or twins, whose polar interplay generates the world. The myths are personifications in ancient terminol-

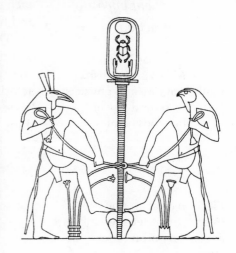

Symbols of the primeval polarities of light and dark that interact to generate the world include the Taoist yin-yang, the Mayan Hunab Ku, and the Mayan glyph signifying "eclipse." Each has a bit of its opposite within itself that drives it endlessly around in search of its complement. The chessboard's alternating dark and light squares represent the field of the world. The Navajo creation myth originates with a primal mother and father, earth and sky. A Hindu story of creation involves an alternation of pulling on opposite ends of a serpent by *devas* ("angels") and *asuras* ("demons"), churning cosmic milk to generate the world. An Egyptian symbol of the nation's unity, called *sma taui*, depicts Hapi, the androgynous symbol of the Nile River, twice. S/he is wearing and pulling on papyrus and lotus flowers, symbols of North and South Egypt, since they grow at opposite ends of the Nile.

ogy of primal polar principles known to modern scientific understanding. Through these stories, the dance of opposites in nature is mirrored in ourselves. These symbols and the works of art employing them are part of larger myths that describe ways to pass constructively through the polarized world.

LOVE AMONG THE ATOMS

Science and technology cannot avoid encountering the ever-present principle of polarity. It is so basic that perhaps instead of teaching science to youngsters in separate pigeon-holes of biology, chemistry, physics, and so on, science courses could investigate the *principles* that run through each of them, such as wholeness, polarity, balance, pattern, and harmony.

The Dyad's fundamental characteristic is the existence of a pair of distinct but equal opposites that seek to unite in an urge to return to unity. Modern scientists know this characteristic as the "laws" of attraction and repulsion. From this basic polarity, light, energy, and matter eventuate. For anything to manifest there must be polarity.

For every action there is an equal but opposite reaction.

—Sir Isaac Newton (1642–1727, English mathematician and physicist, his second law of thermodynamics)

Electron orbit around a hydrogen nucleus resembles a planetary orbit around the sun.

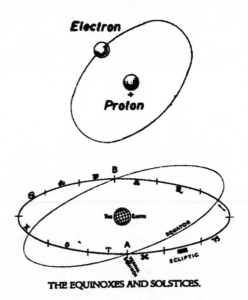

THE EQUINOXES AND SOLSTICES.

The Door of Hydrogen

Like a door from the Monad, the first atom of the ninety-two natural elements to emerge is hydrogen. It is a pure representation of the Dyad, the only element consisting of only two components: a positively charged proton at its center and a negatively charged electron orbiting it as a circumference. All other atoms have a third neutral component (see chapter three). In the role of the Dyad, hydrogen is a door between the unknown unity (Monad) from which it emerged and the subsequent elements ("the Many"), which are built by fusing hydrogen atoms together.

"Static Cling"

Since electrons occur in materials in large numbers, some of them can be loosened by mere friction and induced to move. Scuffling one's feet across a carpet on a dry day loosens electrons, which run up and sit on the surface of the skin. They jump at the first opportunity to unite with an oppositely charged "lover" such as a metal doorknob. The spark is simply the result of a primal urge of parts to return to the whole, of polarity's craving to merge to unity, the original state of the Monad. Even the everyday "static cling" of clothing and plastic wrap illustrates the urge of the Many to return to the state of the One. Scientists call the strength of the "love" between poles their "voltage."

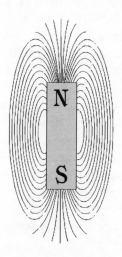

Magnetic field. Magnets are pure expressions of the Dyad. Experiments with them demonstrate its principles: opposite poles attract each other, and alike poles repel. No magnet with other than two poles has ever been found or can be created.

Cell division. The human body grows through the process of mitosis, or cell division, as each cell replicates itself and then divides into two identical "sister" cells. The Monad becomes the Dyad as one becomes two, which becomes four, eight, sixteen, and so on until we have a complete cellular organism: the Many.

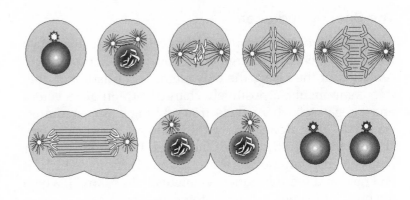

$$1 + 1 > 1 \times 1 \text{—The Unique}$$
$$2 + 2 = 2 \times 2 \text{—The Door}$$
$$\left.\begin{array}{l} 3 + 3 < 3 \times 3 \\ 4 + 4 < 4 \times 4 \\ 5 + 5 < 5 \times 5 \end{array}\right\} \text{The Many}$$

The simplest arithmetic properties of any number reveal its deepest nature. Two is the balance or door linking the One with the Many.

THE BIRTH PORTAL

The study of number-words in many cultures reveals an early awareness that multiplicity in nature and arithmetic comes from the blending of opposites. For example, the ancient Sumerian words for one and two are also those for man and woman.

Today, we consider one and two as merely number quantities, not realizing they are symbols of basic facts of existence. Surprisingly, ancient mathematical philosophers did not consider one and two to be numbers themselves since their representations—point and line—are not actual. A point has no dimension and a line just one dimension. Nobody can hold a true point or line in his hand. Likewise, no one or two points, lines, or angles will create any actual form by themselves. But an ongoing interplay beginning with a point and line is all that's required to construct the world's geometric patterns. Thus, the Monad and Dyad were considered by the ancients to be not numbers but the *parents* of numbers. Their mating, the fusion of the principles of one and two, point and line, unity and difference gives birth to all subsequent archetypal principles revealed as numbers, symbolized by numerals, and seen as shapes in nature. The Dyad, then, is the doorway between the One and the Many.

The ancient mathematical philosophers discovered this by studying arithmetic. They noticed how numbers act when added to and multiplied by themselves. For example, "one" is the only value that, when added to itself, produces a result *greater* than when it is multiplied by itself. One plus one is *more than* one times one.

"Two" is also unique but in a different way. This is the only case where the addition of a number to itself yields the same result as it does multiplying by itself. Two plus two *equals* two times two. Two represents a balancing point between unity and all subsequent numbers, between One and the Many.

The remaining numbers three, four, five, and so on have in common the fact that when each is added to itself the result will always be *less than* when each is multiplied by itself. Three plus three is *less than* three times three. The same for four, five, six, and forever onward: the Many.

Symbolically, "two" acts as an intermediary, a transition, a door or portal between the Monad and all the rest of the numbers. Twoness is the hole or lens through which unity becomes and balances with the Many.

This is the geometric lesson of the two linked circles, symbol of the Dyad. The almond-shaped zone of interpenetration between the circles has attracted the attention of geometers, artists, architects, and mythmakers through history. This is the *vesica piscis*, in Christian cultures a reference to Christ as the "fish" in the Age of Pisces. It's called a *mandorla* ("almond") in India. It was known in the early civilizations of Mesopotamia, Africa, Asia, and elsewhere.

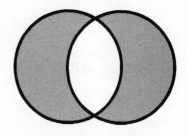

Vesica piscis, or mandorla.

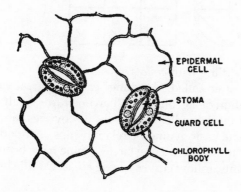

Pairs of "guard" cells on the underside of a leaf contract and relax to form a *vesica piscis*–like opening called a "stoma," allowing the plant to exchange oxygen for carbon dioxide and breathe.

This almond space is the crucible of the creating process. It is an opening to the womb from which geometric forms are born. It brings forth shapes and patterns from the archetypal world of ideal geometry. The *vesica piscis* is a *yoni* (Sanskrit term for the female generative organs) through which the geometric shapes and patterns of our universe emerge. For this reason it has also been called the womb of Chaos, the womb of the Goddess of Night, and the mouth that speaks the word of creation. The *vesica piscis* is where the geometer assists in the birthing process.

Corporate logos are often designed around the *vesica piscis*.

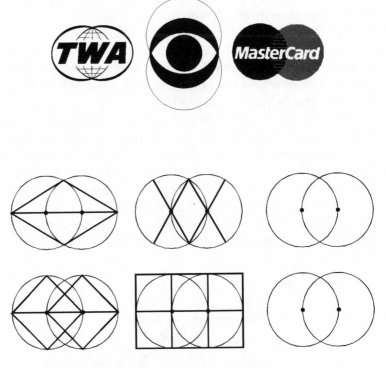

Activity: Doodling with *vesicas pisces*.
    What is often called "doodling" can be of great value to the imagination of a geometer. When one turns off the mind's logical mode and doodles intuitively, discoveries can take place that illuminate basic geometric relationships. So doodle within and around these intersecting circles. Some have been started. Connect points, follow curves, shade areas. Doodle however you like.

Discover the *Vesica Piscis* in Ancient Art and Architecture

The *vesica piscis* was used functionally and symbolically in the construction of doors or portals between mundane and spiritual spaces. Use the geometer's tools to replicate these different styles, representing creeds that are different but share the same geometry.

Architecture aims at Eternity; and therefore, is the only thing incapable of modes and fashions in its principles.

—Sir Christopher Wren
(1632–1723, English architect)

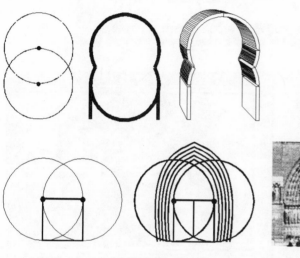

The Moslem arch and cathedral doorway are both designed using two intersecting circles and *vesica piscis*, but at right angles to each other. Through these doors, symbols of spiritual passage, we leave the street of the Many and enter the domain of the One.

The west facade of the Cathedral of Notre Dame in Paris is designed as a circle descending from heaven above which intersects a circle rising from earth below. Entering a cathedral symbolically places us in the space between them where transformation is possible.

Hindu temples are traditionally designed upon geometry originating with the *vesica piscis*. Temples represent a portal between earth and the heavens, our mundane and spiritual identities. The geometry within the circles determines the floor plan of the temple, revealing the detail to which they were planned.

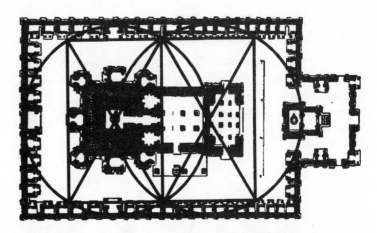

Design using the *vesica piscis* is so basic that we find it underlying human creations thousands of years apart, as seen in this jewelry of the Egyptian pharaoh Tut-Ankh-Amon and the British Royal Seal. Intersecting circles have been drawn within them. To see how their designs stem from the geometry, draw straight lines between the intersections of the circles and their centers, extending them to the opposite sides of the *vesica piscis*.

The Birth of the Many

One pebble tossed into a quiet pond imparts energy we can see as expanding concentric rings. Two pebbles tossed simultaneously a distance apart create intersecting rings and weave a mathematical wonder: concentric, watery *vesicas pisces*. Nodes or points occur everywhere the circles cross. Remarkably, these points establish the precise mathematical distances and relationships required to construct the basic geometric shapes and patterns that recur through the universe. Because all circles are similar to each other regardless of size, the geometric relationships remain constant no matter how large the ripples become.

Since the Pythagoreans considered the first ten numbers to be seed-patterns for all the principles of the cosmos, a geometer need only create their shapes to model all the universal rhythms. The first three shapes to emerge from the *vesica piscis,* the triangle, square, and pentagon, form the only relationships or ratios required to generate all the rest (except for sevenness: see chapter seven). These relationships, called the square roots of two, three, and five, *look* like numbers and are *called* numbers, but they aren't like other numbers. They are like seeds from which the shapes emerge and are nourished. These "root" relationships are expressible not as whole numbers but as never-ending decimals (1.4142135 . . .; 1.7320508 . . .; 2.2360679 . . .). As statements made by the geometry of two intersecting circles and the *vesica piscis,* these ongoing relationships hold the structural patterns for all numbers and shapes that follow.

The "parents" of numbers, symbolized by Monad and Dyad, one and two, point and line, when united as a *vesica piscis* generate the mathematical "root relationships" underlying the shapes and patterns of the natural world. When the distance between centers is one (unit) then the relationships reveal themselves. They're responsible for the basic two-

Oh, grant me my prayer, that I may never lose the touch of the one in the play of the many.

—Rabindranath Tagore
(1861–1941, Indian poet)

dimensional shapes, the triangle, square, and pentagon. When constructed on paper the shapes may be folded to make the five equal divisions of three-dimensional space known as the "Platonic volumes."

Each of these shapes will be constructed and explored in subsequent chapters.

THE WAY THROUGH

In the Dyad we see the Monad refract as Two. The Dyad emphasizes difference. It foreshadows the world's apparent boundaries, conflicts and echoes our own sense of separation. Opposites appear when separateness begins.

Like the line that both connects and separates, the Dyad unites as well as divides. Its representative, the *vesica piscis*, envelops a line as well as indicating its center, creating the appearance of a division but at the same time providing us with the portal through it.

Number proceeds from unity.

—Aristotle (384–322 B.C.)

The opposite poles of the Dyad retain their memory of the One. At every chance they seek to merge and become whole again. Like a lover, lightning leaps a gap to meet and dissolve with its opposite charge where they can both return to a state of unity and peace.

In the metaphor of arithmetic the Dyad reveals itself as the door between the One and the Many, between unity and all other numbers. The *vesica piscis* was recognized and applied in ancient art and architecture as a functional symbol of the fusion of opposites and as a passageway through the world's apparent polarities.

But seeing the Dyad outside ourselves is only half the picture and half the passage. We cannot avoid the polarities within us. The great dual throng of language is evidence of our own investment in a private netherworld of polarized images.

As an exercise, examine the poles within your own thoughts, emotions, and desires. Consider your personal polarities, the situations that strongly attract or repel you, the strong opinions you cherish. Such situations and opinions call us like sirens toward which we unconsciously drift. We rarely resist favoring one pole or another. But only as we

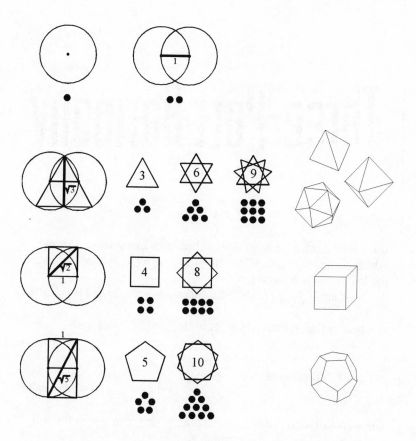

Two intersecting circles contain all the proportions necessary to generate the polygons found in nature and the five Platonic volumes.

realize that the strength we direct toward one is shared by the other can we jimmy loose from both. By finding their common ground, their point of balance, we can pass between them without being pulled toward either side. Only then can we experience the journey with the Monad *through* the *vesica*'s doorway to the further principles underlying cosmic and human design.

Let's continue to use the geometer's tools to assist in the birth of a more recognizable universe. Passing through the Dyad we emerge into the Triad, emissary of the principles of balance and structure.

CHAPTER THREE

TRIAD

Three-Part Harmony

The Triad is the form of the completion of all things.

—Nichomachus of Gerasa (c. 100 A.D., Greek neo-Pythagorean philosopher and mathematician)

We express the principle of threeness by writing the numeral "three" with a beginning, middle, and end.

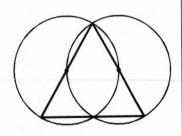

Birth of the Triangle.

But every tension of opposites culminates in a release, out of which comes the "third." In the third, the tension is resolved and the lost unity is restored.
> —*Carl G. Jung (1875–1961, Swiss psychologist and psychiatrist)*

A whole is that which has a beginning, middle and end.
> —*Aristotle*

All good things come in threes.
> —*Folk saying*

The step to Three is the decisive one, which introduces the infinite progression into the number sequence.
> —*Karl Menninger*

With Three a new element appears in the concept of numbers. I—You: The I is still in a state of juxtaposition toward the You, but what lies beyond them, the It, is the Third, the Many, the Universe.
> —*Karl Menninger*

The meeting of two personalities is like the contact of two chemical substances: if there is any reaction, both are transformed.
> —*Carl Jung*

Three is the formula of all creation.
> —*Honoré de Balzac (1799–1850, French novelist)*

The One engenders the Two, the Two engenders the Three and the Three engenders all things.
> —*Tao Te Ch'ing*

ONE, TWO, *THROUGH*

In the legends of King Arthur and the search for the Holy Grail—symbol of self-knowledge and self-perfection—Sir Percival was one of only three knights to actually find the Grail. His name, Percival, derives from "pierce the valley," one who passes safely between looming mountains, piercing polarity, breaking *through* to transcend opposites, showing the way to infinite possibilities where there had only been two.

Without realizing it we "pierce polarity" whenever we count "one, two, three." Counting is a retelling of the original creation myth in the purest archetypal terms. Primitive tribes often count "one, two, many." The Sumerians counted "man, woman, many." But giving separate number names to quantities past two for the purpose of counting is so deceptively simple that we take the idea of a sequence for granted. Yet our ability to name numbers in an ongoing progression reflects a major leap of consciousness. It gives us the ability to transcend polar bounds and realize the unlimited.

Since "one" and "two" were considered by ancient mathematical philosophers to be the *parents* of numbers, then their firstborn, "three," the Greek Triad, is the first and eldest number. Its geometric expression, the equilateral triangle, is the initial shape to emerge through the portal of the *vesica piscis*, the first of the Many.

Speak aloud the word "three" in English or most any language past or present and you'll hear its relation to words like "through" and "threshold" and the prefix *trans* ("across," "penetrate"). The leap to "three," as its linguistic root tells us, takes us over a threshold and *through* past polarized limits of the Dyad. Wherever there are three, as the three knights, three musketeers, three wise men, or three wishes, there is *throughness*, rebirth, transformation, and success.

As the first "number" to succeed its "parents," three is in a unique position within the full sequence of ten. It is the only number of infinitely many to equal the sum of all terms below it $(3 = 2 + 1)$. It is also the only number whose sum with those below equals their product $(1 + 2 + 3 = 1 \times 2 \times 3)$. Mathematically, these three become one and set the stage for the birth of other numbers.

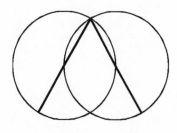

*A*ll *was divided into three.*

—Homer (Ninth–eighth century B.C.)

Examining and understanding tripartite patterns in geometry, myth, religion, and natural science will help us recall our inborn but sleeping memory and our deep understanding of the tripartite world and ourselves.

THREE CHEERS! (NOT TWO OR FOUR)

We can understand the Triad by noticing where it occurs in our lives. It seems natural to divide any whole into three parts. A beginning, middle, and end, birth, life, and death, the three dimensions of space (length, width, height) and time (past, present, future), three daily meals corresponding with the sun's three stages (dawn, noon, dusk), phases of a chess game (beginning, middle, endgame), computer jargon (input, output, throughput), cycle of a traffic light (green—go, yellow—caution, red—stop), and confrontation in the animal and human realms ("flee, fight, or fortify"). The triple-goddess religions of ancient agricultural settlements identified the moon's three phases (waxing, full, waning) with the planting cycle (plant, harvest, dormant), as well as with the goddess herself (virgin, mother, crone). These ("this, that, and the other") are just some of the ways in which we frame and express what we feel to be true about the inherent tripartite wholeness of the universe.

Subdividing each phase into three parts continues the process. People structure beginnings with "on the count of three . . ." "ready, set, go," and "ready, steady, go," cheer their ongoing with "hip, hip, hooray!" or "hurrah!," which derived from "Hu Ra," the Cossack battle cry signifying "Paradise!" We also end events in threes, as in "going once . . . going twice . . . sold!" and baseball's "three strikes, yer out," a game played in First, Second, and Third World countries. We say something is "as easy as one, two, three" but try saying "easy as one, two" and "easy as one, two, three, four." Likewise, we say our ABC's, not our AB's or ABCD's. One, two, four, or any other number of steps don't feel right—do they?—when epitomizing a whole process. Fewer than three feel incomplete, and more than three seem superfluous. These values aren't culturally taught but emerge from deep inside us according to our sensitivity to the archetypes

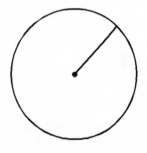

The unity of the circle manifests as a trinity: center, radius, and circumference.

Icons and logos. The triangle is a visually arresting shape and powerful symbol calling for us to be alert. Variations on the triangle have a captivating psychological effect on us that imply a message of wholeness, strength, stability, and process in commerce, science, and religion.

100%
Recyclable

within. Three announces wholeness and completion through an embracing synthesis. We feel its correctness and express it through our words and images.

Three also indicates plurality and supremacy. Drawing three vertical lines | | | and repeating a character thrice were the methods used to indicate "many" and "maximum" in Chinese, and ancient Egyptian hieroglyphics ("thrice-great Djeheuti" [later Hermes Trismegistus]). The Egyptians wrote "thrice dark" to indicate the highest degree of darkness, "thrice hidden" to indicate the unknowable. We cannot help but use three terms to indicate superlatives like "good, better, best," and "fine, finer, finest," honored by the three grades of Olympic medals (Bronze, Silver, Gold).

The superstition about not walking under a ladder derives from a caveat from ancient Egypt: "Don't break a triangle"; that is, don't disturb a complete event.

Some say that "whatever you do comes back to you three times" and "third time lucky." Others say "three on a match" is unlucky. Whatever the value, good or bad, it's three that we speak of. Religious rituals calling for repeating a phrase three times, three prayers, or chanting three times, like superstitiously spitting three times, are intended to make an event happen, to make it whole and bring it to manifestation and completion.

Threeness as *throughness* overwhelmingly recurs in religion, myth, superstition, fairy tales, mysticism, magic, folk sayings, proverbs, and fables, the time-honored expressions that cross cultures.

We can discover more of the message of threeness by making a list from our own experience. In childhood we learn of Goldilocks and the Three Bears ("too hot, too cold, just right"), three blind mice, three little kittens, and three little pigs. In fairy tales and myths we find three wishes, chants, attempts, doors, children, animals, hands, eyes, suitors, wise and weird men and women, hermits, kings, feathers, seeds, sons, brothers, princes, daughters, sisters, princesses, dreams, three-headed dogs, dragons, and other trimorphs, the goddess-trinity, the three Fates (who spin, measure, and cut the thread of each life), three Graces, Furies, and Charities. We read tales of three gifts, days, jewels, rings, goblets, arrows, swords, keys, gates, paths, songs, magic words, rivers, judges, challenges, and tasks. Tridents and triple thunderbolts allude to the greatness of those they accompany. The symbol of the Greek god Apollo was the tripod, or three-legged stool, upon which the respected oracle of Delphi sat and prophesied. The Greek goddess Harmonia (literally "fitting together") was born of parents Aphrodite (Love) and Ares (War). The triple goddess Hecate presided over crossroads in the form of a crone with three bodies and

Alchemical androgyne (literally "man-woman" in Greek) symbolizes transcendence by the union of opposites within us.

three heads. The letters of her name have values that add up to 334, a permissible single unit from 333, emphasizing her threefold nature.

Consistent themes in these stories emerge quickly: opposites are balanced by a third, mediating element that reconciles a conflict, healing the split of polarity and transforming separate parts into a complete and successful whole. A goal is actualized after two failures. Obstacles are overcome on the third trial. The poles of hero and enemy are transformed by the role of a third person, the helper. Triptychs, or three-part paintings, display the three scenes, or stages, of such events.

The high occurrence of ternary rhythms, from wishes granted to divine trinities, is more than coincidence: again a number expresses irresistible archetypal principles from deep within us. We respond to symbols of the archetypes outside us to the extent that we're sensitive to their principles buried within our deeper nature.

THE BIRTH OF THE TRIANGLE

If you're ever responsible for constructing a universe or part of one, you'll surely need to know how to construct a triangle, the calling card of the Triad. The triangle is the first shape a geometer brings through the opening of the *vesica piscis*. In constructing a triangle we witness the birth of a shape from an unspeakable archetypal level into configuration. Every triangle we construct is not merely a copy of another person's triangle but is an approximation of an ideal principle. All the important points needed to construct a tri-

*O*ne whose intelligence has attained to Unity casts away from him both sin and virtue.

—*Bhagavad Gita*

*T*he Triad has a special beauty and fairness beyond all numbers, primarily because it is the very first to make actual the potentialities of the Monad.

—Iamblichus (c. 250–c. 330, Greek Neoplatonic philosopher)

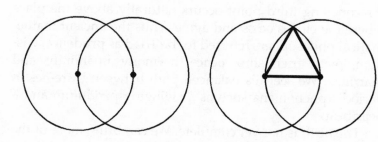

Even the impossible requires a trinity. Three planes drawn at right angles to each other are the minimum required to produce figures like this tribar, the first such design unable to be built in three dimensions.

angle are already apparent in the *vesica piscis;* it only remains for us to connect them.

To make the Triad's principles visible, begin, of course, with a point. From that, create a *vesica piscis.* Connect the circles' centers with each other and with the point above where the circles cross. Now welcome the firstborn of the primeval parents, archetypal opposites, in the form of the triangle (from the Greek *trigon,* "three-sided"). It is the first area enclosed by straight lines, the birth of surface and structure. The emergent triangle remembers its source in the Monad by having equal sides and equal angles: hence, it's referred to as an "equilateral" or "equiangular" triangle.

A triangle is the most astringent shape. In contrast with the circle, which encloses the greatest area within the smallest perimeter, the triangle has the opposite property: it encloses the smallest area for the greatest perimeter. A triangular slice of pizza actually offers fewer toppings than would a round slice with the same perimeter. Recall the experiment in chapter one with a loop of string on the checkered surface. This time verify that a triangle covers fewer squares and a smaller area than any other shape made with the same loop.

The Triad divides the triangle, its emblem, into threes. In all the centuries, geometers have still discovered only three types of triangle: equilateral, in which all three sides and angles are equal; isosceles, in which two sides are equal but the third is different; and scalene, in which no sides or angles are equal. Likewise, all the world's geometry finds only three types of angle: acute, or less than ninety degrees; a right angle equal to ninety degrees; and obtuse, or greater than ninety degrees.

A triangle is a statement about relationship and balance. As the centers of the two circles repel and tug at each other, a reconciling third point occurs naturally above the place where the circles cross and agree. Thus the ancient mathematical philosophers referred to the Triad as prudence, wisdom, piety, friendship, peace, harmony, unanimity, and marriage. We see this balanced path between extremes in periodic phenomena such as a swinging pendulum and a heartbeat.

The birth is not yet complete. We can bring more of the triangle through the *vesica piscis* by extending the lines from

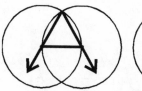 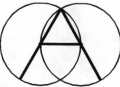 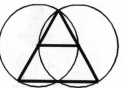 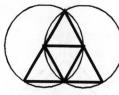

the crossing point down *through* the centers until they reach the opposite sides of the circles. Connect those points with a horizontal line to complete the large triangle. Notice how this line passes precisely through the point where the circles cross below. Connect that point to each circle's center to fill the large triangle with four equilateral triangles. Ancient philosophers were struck by the profound harmony of the *vesica*'s space, which generates triangles by its natural points.

The mathematical relationships between the sides of a triangle and the *vesica piscis*, which gives birth to it, can be made more comprehensible by making them audible. Simply do the construction on a wooden board. Then hammer nails halfway in at the points that make the triangle and at the circles' centers. Tie a single guitar string or piano wire to one nail, and wrap the remainder around each of the nails. Tie the end tightly to the last nail. Pluck the string between each pair of nails to hear a tone. Equal lengths should produce equal tones. Pluck the length of the triangle's height and the distance between the circles' centers in sequence or simultaneously to hear the significance of the mathematician's phrase "the square root of three to one," the seed ratio of the equilateral triangle.

THE ARCH OF EXPERIENCE

Design is but another name for natural law.
  —*Moses ben Maimon (Maimonides) (1135–1204, Jewish philosopher)*

When designing a universe you'll need triangles to create self-supporting structures. The triangle is, in fact, the definition of structure since it is the only polygon structurally rigid by virtue of its geometry alone. It is synergistic in that its stability and superior strength are unpredicted by any of its parts, which, by themselves, do not have these properties.

*S*urface is composed of triangles.

—Plato

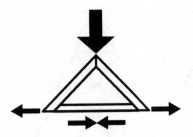

How an architectural A-frame truss supports weight.

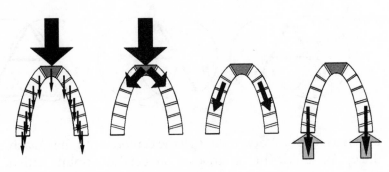

How an arch diverts weight.

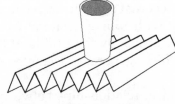

Our pelvis is the keystone supporting our body.

An arch never sleeps.

—Ancient Hindu aphorism

A close look at the fundamental structures of the world shows they're braced by triangles.

Triple-structured architecture, including arches, is fundamental to natural and human designs. Unlike any other shape, the three sides of a triangle resolve opposite tensions into one solid, stable whole needing no support from without. A triangle is self-sufficient. In an arch, the third element, the point atop the triangle where the circles cross, is the keystone that diverts weight downward around its opposite sides into the earth. In a similar way, our pelvis is the keystone atop the triangle formed by our legs. To feel its strength, stand with your legs slightly apart and notice your torso balanced at the pelvic keystone. Architects and structural engineers employ the principle in the omnipresent A-frame truss. Carpenters know to stabilize a structure by bracing it with triangles. We see the same principle in the curve of insect shells, dams, tunnels, bridges, and monuments like the Eiffel Tower. Triangles bestow strength, balance, and efficiency of space, energy, and materials. Fold a paper into an accordion shape and see how triangles support the weight of an object. The more triangles, the more weight it will support.

Triangular structure gives the rose's thorn and shark's tooth their bite, and the wedge and ax their splitting power. Glassblowers have long known that making an arched indentation at the base of a wine bottle strengthens the whole container while also collecting sediment.

We incorporate triangles into our creations for reasons of superior structure, strength, efficiency, balance, visual appeal, and symbolism.

Discover the Tripartite Pattern in Natural Forms

Life-forms in general have a three-part structure. From the body parts of an insect (so named because it's "in sections") to the human body's head-torso-legs with their own subsequent tripartite subdivisions and the three layers of heart muscle, all express the same principle. The geometry of fruits and vegetables will be tripartite when they begin as three-petaled flowers. Observe before you eat them.

If you're going to construct a universe you should also know how to dissolve it. This is done by learning how to create cracks properly. When we look at cracks in a sidewalk we're actually seeing microscopic lines of force magnified millions of times. In nature's view, concrete and rocks, drying mud, and coffee grounds left in the filter are elastic materials. They crack relatively "suddenly" over moments, hours, days, and months along three-way 120-degree joints because this is the way their coequal molecular bonds tear apart rapidly, like an A-frame in reverse. If you have more time—years, decades, or centuries—you'll see materials like old paint and porcelain crack more slowly, resulting in an alternative, four-cornered pattern (see chapter four). Prac-

[An arch is] two weaknesses which together make a strength.

—Leonardo da Vinci

Force without wisdom falls of its own weight.

—Horace (65 B.C.–8 B.C., Roman poet and satirist)

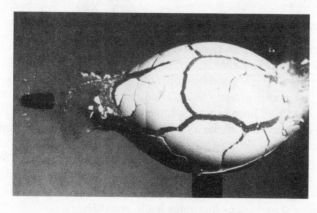

Bullet quickly piercing an egg results in three-cornered cracks.

Sidewalk cracks display three-corner joints.

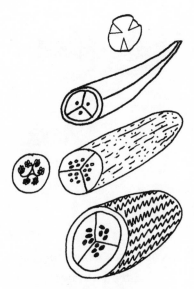

Many fruits and vegetables display three-corner structure.

tice reading the different crack patterns by noticing them in familiar places.

A triangle can be geometrically "decomposed" or subdivided into smaller triangles that diminish by the same root-3 relationship. Follow the instructions below and decompose a triangle. Then superimpose this geometry onto the illustrations of the flowers, fruit, microlife, and insects on the next page to reveal their internal tripartite design structure.

Within each *vesica piscis*, a large triangle has been constructed. Decompose it to see each life-form's design. Some have been done to show you how. Every triangle can be subdivided into four smaller triangles. For example, further subdividing the upper triangle of the fly diagram will show you how its head is formed. Do this construction on the images of other tripartite natural objects you encounter.

T hree is an unfolding of the One to a condition where it can be known— unity becomes recognizable.

—Carl Jung

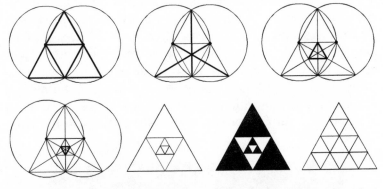

Construct a *vesica piscis* and a large triangle within it. Subdivide the triangle into smaller triangles. Draw three lines connecting each of the triangle's corners with the midpoint of their opposite sides. The points already exist. The lines meet at 120°.

Those three lines cross the innermost triangle at points that can be connected to make a smaller but inverted triangle. Repeat the process connecting each new set of three points where the lines cross each smaller triangle to draw the next triangle. Continue constructing smaller triangles as your sharpened pencil allows. Also, extend the sides of the inner triangles to see how triangular structure perpetuates itself inwardly.

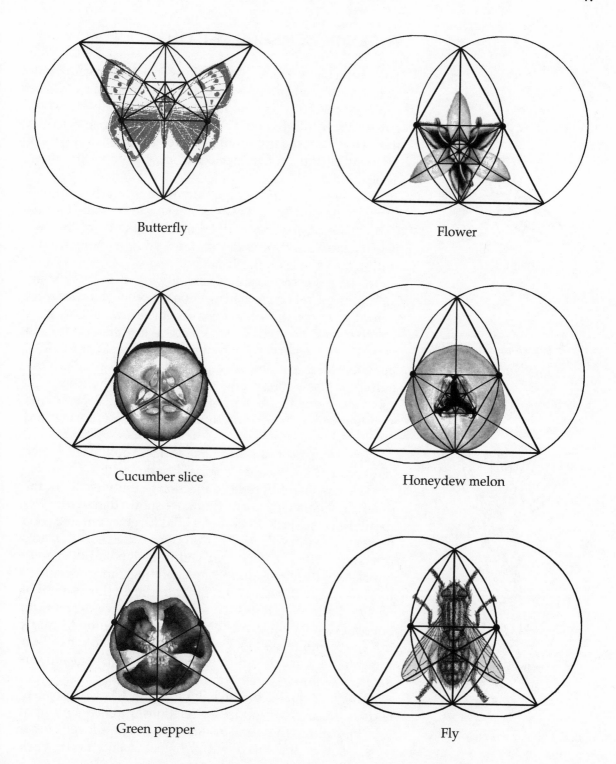

Butterfly

Flower

Cucumber slice

Honeydew melon

Green pepper

Fly

THREE REMEMBERS THE ONE

Trefoil knot

A threefold cord is not quickly broken.

—Bible (Ecclesiastes 4:12)

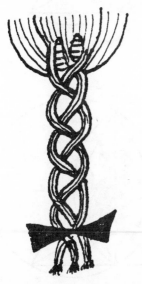

Braided hair weaves a three-in-one whole.

The geometry of knots, with their over-under crossing patterns, is of great interest to modern mathematicians. The simplest possible continuous-line knot, the trefoil knot, crosses itself in three places. No knot can have fewer than three crossings, the minimum required to complete a whole. The word *trinity*, in fact, derives from "tri-unity," or "three as one."

The mystery of three-becomes-one can be studied through the sign known as Borromean Rings, named for its appearance on the crest of the Borromeo family of the Italian Renaissance. It provides an interesting metaphor illustrating the essence of the Triad.

At first glance the sign seems to consist of three interlocked rings. Yet on close inspection we see that *no two rings actually interlock:* the white goes *over* the gray, the gray goes *over* the black, and the black goes *over* the white. Alone or in pairs they're separate. Each maintains its distinct identity. But when all three come together they lock as one, like a trefoil knot, a self-supporting whole beyond what the parts could do alone or in pairs. It takes all three interlocking simultaneously to become unified. Cut any one ring and they'll all come apart.

Construct a set of Borromean Rings. You'll need paper, cardboard or index cards, compass, and pencil.

The Triad's retrograde connection to unity recurs in the braiding of hair. Ordinary experience shows that two tresses twirled around each other won't hold together but will soon unwind unless they're tied at the bottom. That is, the Dyad's principles will not create an enduring whole but merely opposites without resolution, an unending dance of conflict, like lawyers without a judge. But introduce a third strand of hair and alternate the three strands to weave a braid, an ongoing trefoil knot, a self-reinforcing stable unity. A braid's three separate tresses lock as a single new whole like Borromean Rings, needing no outside tie, self-sufficient and stable from within. A bow is only for decoration.

This is a clue to the archetype of the Triad: a properly chosen third factor induces a relationship between opposites that unifies them and brings them to a new level. Threeness resolves the Dyad into wholeness and unity, back

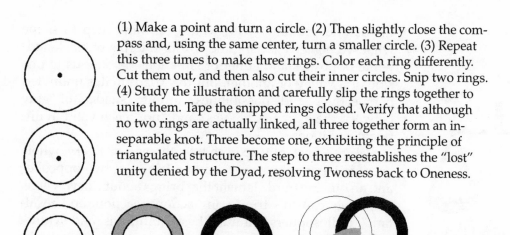

(1) Make a point and turn a circle. (2) Then slightly close the compass and, using the same center, turn a smaller circle. (3) Repeat this three times to make three rings. Color each ring differently. Cut them out, and then also cut their inner circles. Snip two rings. (4) Study the illustration and carefully slip the rings together to unite them. Tape the snipped rings closed. Verify that although no two rings are actually linked, all three together form an inseparable knot. Three become one, exhibiting the principle of triangulated structure. The step to three reestablishes the "lost" unity denied by the Dyad, resolving Twoness back to Oneness.

through to their source in the all-pervading Monad. The embroidered cosmos is braided into existence. In the Native American languages of the Nez Percé and Shahaptian, the number three is known as the "center of the one."

The Triad pervades our lives. The principle of the Triad as seen in the braid, that a minimum of three elements is required to weave into a whole, teaches us how to reconcile conflict. No enduring resolution of any kind is possible without three aspects, two opposites and a neutral, binding, balancing, arbitrating, transforming presence. Knowing how to choose the third factor means the difference between a conflict's resolution and its perpetuation. The third element, if properly chosen, "pierces the valley" and achieves a previously unknown level of experience, one of balance and completion. For example, if I can't decide whether to go out or remain indoors I might consider another factor to induce a decision: the weather, my friends, responsibilities, or needs.

Similarly, we act as relative opposites to some people and have an arbitrating influence between others. If our personal and professional relationships could be visualized graphically, they might appear as a Persian carpet woven of braids within braids.

We can also use an understanding of the Triad to avoid the lure of the dual throng of polarized language. Since the

Lighting six, or twice three, braids of the Havdalah (Hebrew for "Differentiation") candle at the close of the Jewish Sabbath represents the act of discriminating twilight from day and night, and human life woven between the sacred Sabbath and secular weekdays.

words and grammar with which we think help to shape what we perceive, seeing the world in terms of black-and-white opposites and either-or situations detours us to one side of a valley or another. But the Triad invites unlimited perspectives and possibilities, infinite shades of gray between the extremes of "yes" and "no," each valid in different situations.

We're constantly exposed to threeness. Every whole event is inherently comprised of a trinity of two opposites and an outside third element that brings about a new whole. Physicists call this trinity an "action, reaction, and resultant"; philosophers call it a "thesis, antithesis, and synthesis." The three elements together form a greater new thesis, which, in turn, induces its opposite and is ready for a greater synthesis.

In becoming whole the Triad, threeness, remembers unity and expresses its yearning to return to the state of the Monad.

The following examples each comprise a triune whole, a Borromean knot or triple braid. Use them to learn to recognize the three components in any whole event. Begin by dis-

The principle of the Triad is summed up in a simple hanging balance: two opposites are reconciled by the presence of an independent third aspect above them both. This is why scales symbolize justice by *trial*.

OPPOSITE POLES		BINDING ELEMENT	NEW WHOLE
Seller	Buyer	Money	Commerce
Man	Woman	Love	Home
Water	Flour	Fire	Bread
Proton	Electron	Neutron	Atom
Prosecution	Defense	Judge	*Trial*
Cook	Eaters	Food	Meal
Artist	Medium	Inspiration	Art
Projector	Screen	Film	Movie
Management	Labor	Arbitrator	Mediation
Wick	Wax	Flame	Candle
"North" Pole	"South" Pole	Field	Magnet
Negative	Positive	Wire	Electric Circuit
Reactant	Reagent	Catalyst	Chemical Process
Left Pan	Right Pan	Central Hinge	Balance Scales

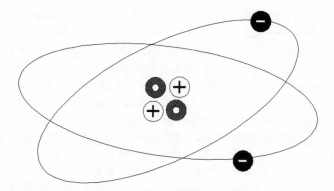

Helium, the second atom to manifest, the first element composed of three different "charges."

This world is a vast unbroken totality, A deep solidarity joins its contrary powers.

—Sri Aurobindo Ghose

tinguishing the two opposites, and then note the catalyst or binding element that causes the new relationship to arise.

Tripartite architecture braids the cosmic tapestry. Albert Einstein's famous equation $E=mc^2$ describes the whole cosmos as Borromean Rings of three distinct but unified aspects of energy, mass, and light. Modern mathematical physicists describe mass, commonly called matter, as atomic whirlpools. After the emergence through the polar portal of manifestation, all subsequent atoms beyond hydrogen are composed of three "parts," like spinning Borromean Rings of positive protons and negative electrons balanced by neutral neutrons. Atoms configure in patterns along lines of force whose surface features we smooth out and simply call "things."

Color is also based on threes. Direct light emanates from an illuminating source like the sun, a star, or television set. Pure whole white sunlight is itself a threefold weave of primary colors (red, green, blue-violet). Light's primary colors are "additive" because their combinations braid together to recompose whole white light. Verify this by shining three flashlights, each with a different primary colored filter, at one spot on a wall to produce white. Look closely at a color TV screen with a magnifying glass and you'll see the three primary color cells whose light blends in our eye to form the colors on the tube's surface.

Pigment colors such as those in paint, ink, pencil, crayons, and chalk occur by "reflected light." Their three pri-

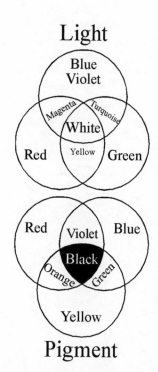

Light

Pigment

Colors of light and colors of pigment build in opposite directions.

Shinto symbol of the tripartite revolving universe.

maries (red, yellow, blue) are called "subtractive" colors because they *remain* after direct light colors are absorbed into or "subtracted" from the surface. Just as direct light adds toward white, pigment colors subtract toward black, the absence of color. The reflected colors of the objects around us show us exactly what the actual color *isn't* because its true colors were absorbed. What reflects to us is what remains. The ink on this page is black because all light is absorbed into it or subtracted and what remains is no color at all.

The principle of the triple weave governing colorful light vibrations occurs also in sound and music. It takes combinations of only three primary notes (the tone, fourth, and fifth) to combine in four secondary ways to produce the natural seven-note musical scale of a vibrating string or column of air (see chapter seven). The triple weave is the basic knot of this embroidered universe at each level of matter, energy, and light.

The fact that human body structure is three-part should be no surprise since the body begins life as a microscopic three-part structure. At three weeks, the human embryo has become a sixty-cell enfolded sphere called a *blastula* (Greek for "bud") that enfolds and diversifies into three distinct layers to become the different but intertwined systems of our body.

We may think of our skeleton as our body's solid unchanging core of bone. It shouldn't change much, should it? But like other living cells it embodies a dynamic three-phase cycle of transformation: emergence, development,

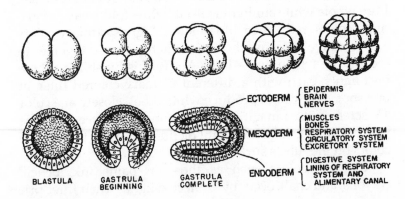

Embryological development of a vertebrate animal.

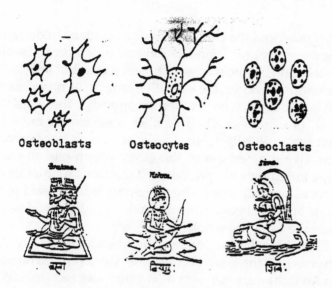

Osteoblasts   Osteocytes   Osteoclasts

The three types of bone cell correspond with the three Hindu deities, Brahma, Vishnu, and Shiva, perpetually creating, maintaining, and destroying.

and dissolution, or, simply, birth, growth, and decay. Thus there are three types of bone cell. The *osteoblast* (Greek for "bone bud") creates new bone and repairs old ones by extracting calcium from the blood and laying down a bone matrix. *Osteocytes* ("bone cells") maintain and repair the bone's substance and strength. And *osteoclasts* ("bone breakers") dissolve and reassimilate old bone materials by dissolving bone matrix into the blood. Osteoblasts reabsorb it to make new cells, and the cycle continues.

The ancient trinity of gods of the Hindu religion called *Trimurti* (Sanskrit for "one whole having three forms") mythically personifies the functions of the threefold bone cycle: Brahma the Creator, Vishnu the Preserver, and Shiva the Destroyer. Myth sometimes cannot help but be science in disguise.

The same principle of tripartite unity was consciously intended in the creation of the government of the United States. When its founders were deciding how to construct a stable yet dynamic and enduring system they looked to their background as Freemasons. Esoteric Freemasonry of the eighteenth century used architectural symbolism to convey ancient teachings of self-knowledge and inner development passed down from Egyptian times through a chain that included "Peter Gower" (a thinly disguised name for

I assure you I had rather excel others in the knowledge of what is excellent than in the extent of my power and domain.

—Alexander the Great
(356–323 B.C.)

*T*hou, nature, art my
goddess; to thy laws my
services are bound.

—William Shakespeare

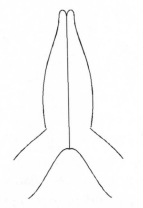

Prayer of 120° forms a trinity.

Pythagoras) and the Gothic cathedral builders. In designing the government the founders devised a democratic system of checks and balances among three separate but interlocked Borromean branches: the executive branch, or presidency, is *intended* to clash with its eternal opposite, the legislative branch of Congress. Their conflict is supposed to be balanced by an ideally neutral judicial branch, the Supreme Court, like the neutron of the atom, which has no biased charge but allows the protons and electrons to interact creatively as a whole atom. The three-part federal braid recurs in state and local governments, each a smaller braid woven within the tripartite whole.

Nature teaches, technology builds, and we know deep within us that three can become one. We accept this concept without thinking when we sit on a three-legged stool. What two legs cannot have done, a third permits. Three brings a process to completion. But this achievement is not necessarily static. It can set into motion a triangular cycle, an ongoing rhythm of birth, growth, and dissolution, a three-phase process with many faces in which all change must partake. Whether an arch transmits weight into the earth or bone cells change along a cycle—and even in the greatest cosmic transformations of light, energy, and mass—the weave is threefold.

TRIUNE SELF

The triangle is the world's preeminent symbol of divinity. As the symbol of holy trinities, it reaches into nearly every religious tradition. In prayer our hands spontaneously form a three-sided 120-degree corner. The urge of the Triad is very powerful because it is rooted in our own triune nature, which is modeled on the world's.

Long before Einstein phrased $E=mc^2$, ancient geometers represented the trinity of light, energy, and mass by their three tools: the compass with its unsleeping eye or sun above and its legs as rays of wisdom and beauty shining into our lives, the straightedge that directs energy patterns of tension and movement, and the pencil and paper that make the patterns visible.

These three tools of light, energy, and mass by which the

Divine geometer constructs the cosmos and by which the symbolic geometer approximates archetypal patterns are also mirrored in us. What scientists call "light, energy, and mass" are the traditional "spirit, soul, and body" described by Plutarch as *nous* (divine intellect), *psyche* (soul), and *soma* (body). To the Hindus the principles within light, energy, and mass are the three *gunas* ("qualities") of purity, activity, and inertia that blend in different proportions in all processes and events outside and within us.

Our "spirit" may be thought of as direct light, the wisdom and beauty of our deeper Self, whose knowing is reflected on the waters of our psyche as our ability to think. What is deeper within us as delight is reflected as emotion. What is deepest within us as the most mysterious "Will to Be" manifests as our most explicit structure, the body with which we act.

Freemasons continued this geometric architectural symbolism in the new government. The United States is one of a few countries whose official seal has two sides, both on the dollar bill. The pyramid on the reverse, with its luminous triangular eye and trapezoidal base below, is an abstract model of ourselves. The eye represents our ever-awake, deep, divine spiritual Self, the thick foundation below its outward expressions, our energetic psyche and denser body.

Note the trapezoidal shape of the base. It's similar to a design often seen as an Egyptian sphinx's or pharaoh's *nemes* headdress, whose slanted sides, like the base of the pyramid, imply an unseen triangle above it representing our little-suspected divine nature.

Saints have halos of different shapes, but only God's halo is traditionally a triangle.

DISCOVER THE TRIANGULAR PATTERN IN EGYPTIAN JEWELRY

The ancient Egyptian priest-craftsmen used geometry extensively in their designs, both structurally and symbolically. Simultaneous representations of Self and the world's structure were often incorporated into their art, crafts, and architecture to remind the viewer initiated in its system of symbols of their significance.

Eighteenth-century French ornament.

Egyptian blue faience bowl. A point in a circle, the hieroglyph for the sun and light, symbolizes the supreme spiritual principle. It sits at the center of a divine triangle as the common eye of three fish swimming in the waters of our psyche. As in Hinduism and Buddhism, the lotus implies unfolding awareness. This bowl was an everyday reminder of the divine triune light at our center.

The trapezoidal shape appears often in Egyptian jewelry. This pectoral of Senwosret II (c. 1890 B.C.) was designed as a trapezoid to fit perfectly at the bottom of an equilateral triangle constructed within a *vesica piscis*. In typically subtle Egyptian symbolism, what is not visible and beyond name or form is best represented by empty or "negative" space, in this case the missing "top" above the trapezoid. Unseen divinity is implied by the "empty" triangular space above the visible form.

A geometric construction superimposed upon the image will help us see how it was designed. Notice how the elements of the composition follow the prominent points, lines, and areas of the construction. The center of the whole triangle is precisely at the tail of the scarab, where the sun or spiritual principle is born. To an initiate who can read its symbols, this pectoral is both a cosmological diagram and a map of our Self. Glyphs of life, time, and eternity are encompassed within the divine triangle. This jewelry, like much of Egyptian art, honors the divine space within each of us.

Further explore the rich geometry of this pectoral by extending lines and connecting obvious points.

○

The principles of the Triad are omnipresent. Taking us past the oscillations of polarity by synthesizing a new

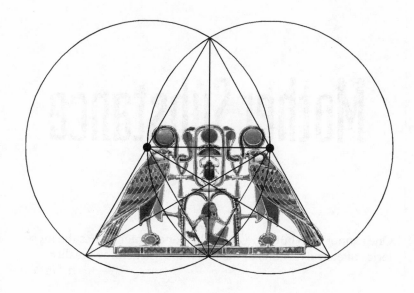

whole, the function of balance appears, natural forms become stable structures, and sequentially ordered processes actualize. We'll see this trinity of "structure-function-order" and "space-power-time" also expanded in the principles of numbers that are multiples of three, in chapters six and nine.

Religion, myth, folk tales, science and art tenaciously reiterate the importance of trinities due to their deep roots within every part of us. They have powerful psychological effects on us because the tripartite universe connects with its archetypal root within us. Threeness shows us that we're not separate from the rest of the universe but are literally braided into it; we are a complete whole living in a greater complete whole.

You already knew all this but may have had to be reminded.

Now having the ability to express the geometric constructions of the Monad, Dyad, and Triad, we can represent the point, the line, and the surface. The principles of numbers next lead us onward to give our universe depth.

Going, going, *gone.*

Pure essence, and pure matter, and the two joined into one were shot forth without flaw, like three bright arrows from a three-string bow.

—Dante

CHAPTER FOUR

TETRAD

Mother Substance

One cannot help but be in awe when one contemplates the mysteries of eternity, of life, of the marvelous structure of reality.
—*Albert Einstein*

. . . to know the order of nature, and regard the universe as orderly is the highest function of the mind.
—*Baruch Spinoza (1632–1677, Dutch philosopher)*

The perfection of mathematical beauty is such . . . that whatsoever is most beautiful and regular is also found to be most useful and excellent.
—*Sir D'Arcy Wentworth Thompson*

In all their works [the ancients] proceeded on definite principles of fitness and in ways derived from the Truth of Nature. Thus they reached perfection, approving only those things which, if challenged, can be explained on the grounds of truth.
—*Vitruvius (First century B.C., Roman architect, engineer, and writer)*

What canst thou see elsewhere which thou canst not see here? Behold the heaven and the earth and all the elements; for of these are all things created.
—*Thomas à Kempis (1380–1471, German ecclesiastic and writer)*

We carry within us the wonders we seek without us.
—*Sir Thomas Browne*

One race there is of men, one of gods, but from one mother we both draw our breath.
—*Pindar (c. 522–438 B.C., Greek lyric poet)*

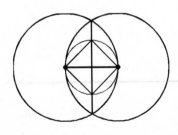

The Birth of the Square.

IN THE SEARCH OF THE DEPTH

Although the ten mathematical archetypes exist simultaneously, it can help to view them as a sequence. We've seen the creating process emerge from the Monad's point to the Dyad's line and Triad's surface, that is, from zero dimensions to one dimension to two dimensions. Our journey corresponds with the emergent structure of a plant: the point of the seed, the line of the stem, the surface of the leaf. We now come to three dimensions, to volume, to the fruit of our geometric plant.

Three points define a flat surface, but it takes a fourth to define depth. The archetype of fourness, called Tetrad by the Greek mathematical philosophers, is expressed in geometry and nature as *volume*, three-dimensional space. The Tetrad provides the framework of anything we can point to.

Like other mathematical archetypes, the principles of the Tetrad are born as shapes through the portal of the *vesica piscis*. But now triplets emerge: the tetrahedron, the square and the cube. The flat square or *tetragon* (Greek for "four sides") has four equal sides. The *tetrahedron* (Greek for "four faces") is a volume of four identical triangular faces. And the cube or *hexahedron* ("six faces"), the most well known volume, is enclosed by six square faces. These forms recur in natural and human creations and symbols.

The geometry of the Tetrad manifests itself in nature with greatest exactness at the borderline between nonliving and living forms. The three geometric forms are widespread as the remarkable submicroscopic structures of atoms, molecules, crystals, viruses, and living cells.

Let's see how the Triad's flat surface blossoms upward to become the Tetrad's volume, how the triangle becomes the tetrahedron, the first fruit of the creating process.

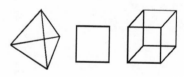

THE BIRTH OF DEPTH

Construct the Tetrahedron

Here's how to bring depth to the universe.

(1) First construct a *vesica piscis* and large triangle. Subdivide the triangle into four smaller triangles by connecting the circles' centers and their intersections. (2) Draw tabs

*H*ast thou entered into
the springs of the sea?
or hast thou walked in the
search of the depth?

—Bible (Psalms 38:16)

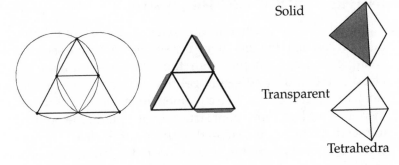

Solid

Transparent

Tetrahedra

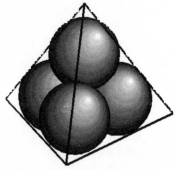

Four spheres stack as a tetrahedron.

extending along the three outer triangles as shown for gluing. Cut out the form around the tabs. (3) Crease each line and fold the three corners upward until they meet at one point. Glue the tabs along the inside of each edge.

Hold the tetrahedron and rotate it. It appears similar from every direction. The tetrahedron is the only three-dimensional shape whose corners are the same distance from each other. Other than the sphere, the universe doesn't allow any volume to have fewer than four corners, or faces. Press it between your two hands, between a corner and its opposite face, to feel its strength. Of all structures, this first three-dimensional shape is the most spare. Space is desiccated to its minimum volume. Buckminster Fuller considered the tetrahedron to be the basic unit of space. While the sphere encloses the most volume in the least surface area, the tetrahedron encloses the least volume with the greatest relative surface area. Being the minimal solid, it is the strongest and most stable.

The birth of the tetrahedron, the first volume, brings the Triad's principle of triangular stability to three dimensions using the fewest materials. We find the Triad in tripods, music stands, three-legged stools, campfires, and wheelbarrows. The tricycle's three wheels form a triangle on the ground to support the seat above with greatest stability. The occasional tetrahedral packaging of milk and juice takes advantage of the volume's great strength.

Insects make use of tetrahedral stability for walking. Having six legs, they walk on three at a time, alternating every other leg from one side to the other. They move forward as if on tripods. The lotus position of yoga stabilizes the body into a self-supporting pedestal for the mind's undistracted meditation.

Insects walk on three legs at a time.

There are always four ways to look at any three-dimensional structure: as points, lines, areas, and volume, or as corners, edges, faces, and from the center outward. This construction built a tetrahedron of four faces. We can also think of it as four corners, made by simply stacking four spheres like oranges whose centers make the corners of a tetrahedron.

In the 1870s the Belgian physicist Joseph Antoine Plateau constructed a tetrahedral wire frame and dipped it in soap solution. To his amazement the film did not just cover the faces but met inwardly to create a fascinating internal structure. After measuring it he discovered that the film followed the path that made it cover the minimal surface area possible yet connect every edge. When he then captured a single bubble in it he was further astonished to see a small, curved tetrahedron at its center. Wire frames are a way for nature to reveal her own minimalist geometry, the one that shows up as microscopic forms, the structures of molecules, crystals, and microlife. Architects have found that by building along the lines of force shown by Plateau's soap films they can create extremely thin yet strong structures that resists compression and tension.

Bend wire to form a tetrahedron and extend one edge to use as a handle. Repeat Plateau's experiments to see how form in nature is a graph of forces.

A thing may endure in nature if it is duly proportioned to its necessity.

—Fra Luca Pacioli (1445?–1514?, Italian mathematician, writer, and math teacher of Leonardo da Vinci)

Nature uses as little as possible of anything.

—Johannes Kepler (1571–1630, German astronomer)

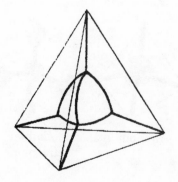

A wire tetrahedral frame dipped in soap solution reveals the geometry of minimal forces, minimal volume, and maximum strength, expressed in nature as the silicon skeleton of this microscopic Nesselarian.

○

The tetrahedron commonly occurs in organic and inorganic chemistry and in the world's submicroscopic structures. Its geometry frames the architecture of many elementary molecules, including methane (CH_4), ethane (C_2H_6) and ammonium (NH_3), the basis of amino acids, the building blocks of life. In each of these three molecules, a carbon or nitrogen atom sits at the center of a tetrahedron at whose four corners are smaller hydrogen atoms. The similar charges of hydrogen atoms' electron pairs cause them to go as far as they can from each other. The result is a tetrahedron, the only three-dimensional shape whose every corner is the same distance from every other corner.

Substances may be composed of identical atoms whose different geometric configurations give them different characteristics and properties. Glass and quartz, for example, are both composed of sand, silicon dioxide (SiO_2). Glass is brittle and transmits visible light while blocking ultraviolet rays. The unstructured molecules of glass flow as a liquid; in fact, glass, the oldest synthetic polymer, is not considered a solid but a supercooled liquid flowing slowly over decades. (Notice how the bottom of an old window settles in thick, wavy lines.) But the same SiO_2 molecules in quartz link so that every silicon atom is at the center of a tetrahedron with oxygen atoms at its four corners. Its extended tripod structure makes quartz a hard crystal, which, like glass, transmits light, including ultraviolet frequencies.

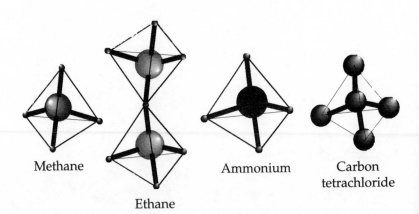

Methane

Ethane

Ammonium

Carbon
tetrachloride

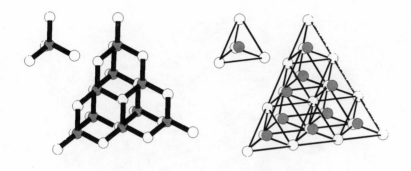

Two views of a diamond showing its carbon atoms linked as a tetrahedral structure. The short bonds between atoms insure the diamond's great strength.

But quartz is not as hard as diamond, the world's hardest naturally occurring substance and best conductor of heat. The diamond is also structured as a tetrahedral network, but only of carbon atoms, each of which sits at the center of a tetrahedron having four other carbon atoms at its corners. Because the energy bonds between the atoms are so short and strong, lying along the edges of a minimal tetrahedron, a diamond is firm and difficult to cut. But diamond cutters know how to overcome its hardness by slicing along the tetrahedral faces.

With the Tetrad come volume and structure. But why is the number four associated with physical manifestation? To better understand the Tetrad let's examine its expression as the square.

BACK TO SQUARE ONE

The square is an obvious symbol of fourness. Unconscious knowledge of it is so strong within us that it wells up in many of our sayings. We speak of getting a "square deal" and a "square meal," of "square living," being "fair and square," "facing problems squarely," and sometimes going "back to square one" in order to "square our accounts." If an event does not "square with our sense of justice," or a story is "not squaring with the facts" we sometimes must have "square-jawed determination" and "stand foursquare," having our feet planted "squarely on the ground." Like its sibling, the tetrahedron, the square is associated with equality, reliability, fairness, firmness, and solidity. "Four, firm and fair" goes the saying.

Numbers, furthermore, as archetypal structural constants of the collective unconscious, possess a dynamic, active aspect which is especially important to keep in mind. It is not what we can do with numbers but what they do to our consciousness that is essential.

—Marie-Louise von Franz (Jungian psychologist and writer)

*I have not kept my square,
but that to come
Shall be done by the rule.*

—William Shakespeare

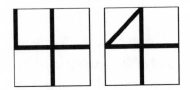

We write the numeral four as
if in 2x2 grid.

*It is hard to be truly
excellent, four-square in
hand and foot and mind,
formed without blemish.*

—Simonides (556–468 B.C.,
Greek poet)

The ancient Greeks noticed that four is the first number formed by the addition and multiplication of equals (4 = 2 + 2 = 2 × 2) and so was considered the first even number and first "female" number. (Even numbers were considered female, odd numbers male. In Hebrew and other languages, the names of parts of the body that occur in pairs, like eyes and hands, have a feminine ending. Words for parts of the body that occur singly, like the nose, have masculine gender.) To the Pythagoreans the even-sided square represented "justice" because four is the first product of equal values, a number divisible every way into equal parts. That is, 4 = 2 + 2 and the two twos can be further divided equally (1+1) + (1+1) and each divided again finally to 1+1+1+1. Four is the first number to reach back to its source in the Monad this way.

The planet earth below us, solid ground, terra firma, is the supreme symbol for substance, mass, volume, strength, and stability. We organize space on the ground by the four cardinal directions of the compass and our body, dividing the circle of the horizon around us into four quarters in front and back of us, to the left and right. The four-cornered cross in a circle has long been the astronomical symbol for planet Earth. The four traditional winds blow across the four corners of the globe. We quarter not only space but time, naturally dividing the year into four seasons based on the relationship of the sun with the earth around the two equinoxes and two solstices.

Fourness indicates association with the stable, solid earth. The ancient Egyptians symbolized this by showing four pillars arising from earth to support heaven. The Mayans likewise portrayed four beings as supporting the celestial roof.

Regardless of the culture, the square was the preeminent symbol for the ancient Earth Mother goddess. The association of the earth with the number four, femaleness, and justice is very ancient, far preceding recorded history. The principles of the Tetrad describe her nourishing aspect: she gives birth, clothes her creations with material substance, and encourages their growth equally. The word *nature* comes from the Latin for "birth." In the Navajo language, nature is called "Changing Woman." The Navajo term for summer, then, is "Changing Woman's happiness." *Mater*, the Latin word for "mother," has given rise to the word "matter," also related to *meter* ("measure") and *matrix*. The Latin word for father, *pater*, gave rise to "patron" and "pattern." Thus, natural forms can be seen as coming from both a mother and father, the mating of matter and pattern.

In the way that a mother provides nourishment for her developing child, the "world mother" is described in myths as nourishing the emergent archetypal patterns with matter, consisting of four types of food, called the four ancient elements: earth, water, air, and fire. These elements provided the mythopoeic way of referring to the modern scientific four states of matter: solids, liquids, gases, and plasma or electronic incandescence. The three denser states are familiar to everyone as ice, water, and steam. The fourth state, which the ancients called "fire," is known today as "plasma," the glowing electrified gases that "burn" in the sun and stars and cause the fiery glow within fluorescent lights and neon signs. We laugh at the simplicity of the ancient concept of just four "elements" comprising the world when today we recognize over one hundred varieties of atom. But no matter who looks at the world or when, we can find only four *phases* of *mater*, four clothings of nature.

Nature unfolds these four states in a particular order. Ice must pass through its phase as water before it becomes steam. If you ever forget the correct order Plato gives in his *Timaeus*, just fill a jar with water, and soil, and air, and shake it up. The natural layering into which it settles will show you the order of the elements, or states: most dense earth on the bottom, then water, above that air, and, most ephemeral, light passing through.

The Japanese glyph for "country" is a square "earth" symbol enclosing the ideograms for "people" and "protection."

Ancient Vietnamese "four perfections" coin symbolizes the square "earth" within the circle of "heaven."

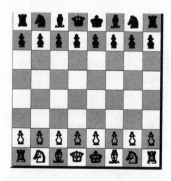

This square chessboard represents material interplay in the world.

*T*he chessboard is the
world, the pieces are the
phenomena of the universe,
and the rules of the game
are what we call the laws of
Nature.

—Thomas Henry Huxley

Being so close to the earth, we express the archetypal symbolism of the Tetrad in many ways without realizing it. In cities we walk on sidewalks inscribed with squares that symbolize the earth they cover. As communities grow they are often laid out by city planners on a "rational" square-grid street pattern. We build right-angled architecture and erect chain-link fences, made of tilted squares, as symbolic walls of earth. Beds, tables, chairs, and thrones are miniature models of the earth. The square insists on serving as a material boundary.

Gridded graph paper is a symbol of the "four corners" of the earth, the largest drawing surface there is. Farmers plow square fields on the traditional forty acres of land. The ancients noticed that four-footed animals are earthbound. Four shows up whenever there is reference to earth and matter.

A baseball "diamond" is actually a square, as is a boxing "ring" where fighters "square off." Many sports fields are square-cornered, signifying arenas of earthly action, material conquest, and physical interplay. The history of board games is full of square-cornered fields of play symbolizing initial equality in the earthly sphere of action, mass, and force.

In classical origami, the Japanese art of paper folding, each construction always begins with a square. Thus, origami cranes, horses, frogs, plants, and people all share a common *mater*. The harlequin's costume with its checkerboard pattern symbolize his role as a chthonic force. There are four traditional colors in the tarot deck (white, black, red, green). All quaternaries indicate the material and perceptible.

Words containing the letters "qu" often hint about their

Hopi symbols for Mother Earth. Labyrinth at Chartres Cathedral, and Italian Renaissance garden plan.

Square labyrinths and those built around squares and crosses are metaphors of our wanderings through the earth, symbolically within ourself, and our path of transcendence.

relation to the word "square" and its Latin counterpart "quadratus" as in quarter, quartet, quart (one fourth of a gallon), quarry (where square-cornered blocks are cut), quarantine (forty traditional days of isolation), and squadron (a group of four).

In the classic symbolism of myth and religion, the number forty (= 4 × 10) marks a passing beyond (see chapter ten) a worldly or fourfold material phase. This symbol of passage lends significance to Noah's rain of forty days and nights; it is also reflected in the life of Moses, whose 120 years encompassed three forty-year phases and who waited

The Egyptian glyph for "town" is an X in a circle, followed by a half circle that denotes the feminine gender. Today we find the same glyph, the X in a circle, in street signs saying "No Standing." In both examples "X marks the spot," this ground, this earth.

nu-t ⊗ , ⊗ , ⊗ , village, hamlet, town, city, community, settlement; plur. ⊗ , ⊗|

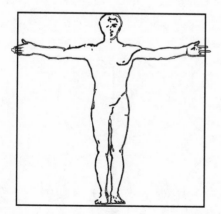

Four basic "key" designs of classic architectural ornament indicate the four ancient elements.

Ape Factor. Mountaineers quantify the extent of their reach by the term "ape factor," the ratio of arm span to height. Since, on average, our height equals the distance of our outstretched arms, an ideal body fits within a square. Its diagonals cross at our genitals, indicating the square's relationship to fertility, birth, and generation. If arm span and height are equal, the ape factor is unity. A higher ape factor means a greater reach, desirable for those who climb and hug the earth.

Earth's four-cornered cracks. When you see three-cornered, 120° cracks in elastic materials, you know that the universe broke quickly there. But when you see four-cornered crossings, 90° cracks in inelastic materials such as old paint, tree bark, porcelain sinks, tiles, and the inside of glazed ceramic mugs, you can be sure that they split slowly over years, decades, and centuries. Whereas the molecular bonds in elastic materials continually adjust their strain in all directions, inelastic materials do not yield so easily, releasing their tension slowly in one direction. The longest cracks are the oldest. Four-cornered cracks are the crow's feet of matter.

forty days on Mount Sinai to receive the Ten Commandments. The Israelites spent forty years wandering in the desert. Jesus' forty days in the wilderness, the forty days of Lent, and Ali Baba's forty thieves each recall the transformation of earth and self, often through physical ordeal. At the fortieth day of human pregnancy, the embryo becomes a fetus.

THE BIRTH OF THE SQUARE

While the triangle was constructed by connecting existing points within the *vesica piscis*, the construction of a square requires the faculty of rational thought: one goes through a sequential, logical procedure in order to discover the hidden points to connect. (Some people are called "squares" and "blockheads" for being too rational.)

There are many approaches to constructing a square. Here are two useful ways:

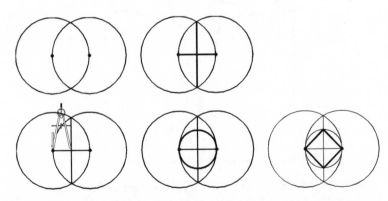

The first is the simplest and most elegant. It arises completely within a *vesica piscis*. (1) Construct a *vesica piscis*. (2) Draw a horizontal and vertical line connecting the centers of the circles and their intersections. (3) Put the compass point where the lines cross at the center of the *vesica piscis* and the pencil at one circle's center. (4) Turn a small circle. (5) Connect the four points where the circle crosses the lines to make the tilted square visible.

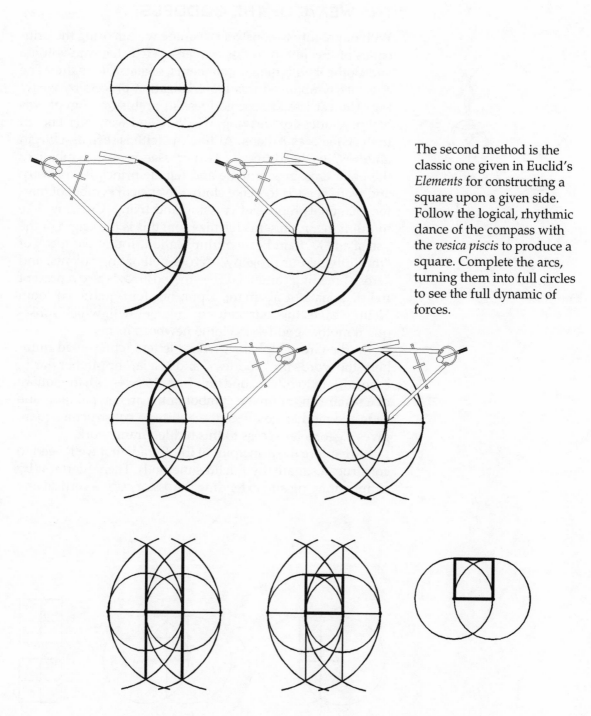

The second method is the classic one given in Euclid's *Elements* for constructing a square upon a given side. Follow the logical, rhythmic dance of the compass with the *vesica piscis* to produce a square. Complete the arcs, turning them into full circles to see the full dynamic of forces.

THE WEAVE OF THE GODDESS

With our ability to construct a square we can bring the principles of the Tetrad to our universe. We can create volume and clothe it with the four elements, or states, of matter. The ancients symbolized this materialization process by weaving. The earliest known goddess in prehistoric Egypt was Neith, goddess of weaving, wisdom, and war, later known to the Greeks as Athena. Although Neith is seen as a virgin goddess, her headdress of two crossed arrows symbolizes the polar crossings of male and female principles, of warp and weft, needed to make cloth, the ancient symbol of matter. Diagonals represent crossing and fixing. Crossing, like multiplying, represents fertilization; this is the origin of the use of an "X" to indicate multiplication. Today, we speak of "multiplying our progeny," "cross-pollinating" plants, and "cross-breeding" animals. The crossed arrows also represent the diagonals of a square. Upon her four-quartered loom Neith weaves the four states of "mater" with which subsequent mother goddesses clothe newborn forms.

In the Greek alphabet, whose letters represented numbers, the words for "goddess" and "Gaia," or Mother Earth, each add up to fifteen, one unit from 16 (= 4 × 4), the square of a square and ultimate symbol of the archetypal feminine as fertile mother goddess; the ability to construct a square gives a geometer access to this hidden framework.

To explore the principles of the Tetrad, first we'll need to construct a square (by Euclid's method). Then quarter it by simply drawing and extending the *vesica piscis*'s vertical line

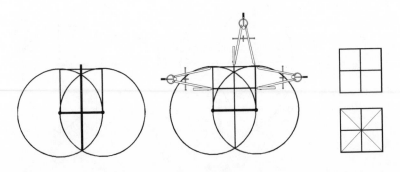

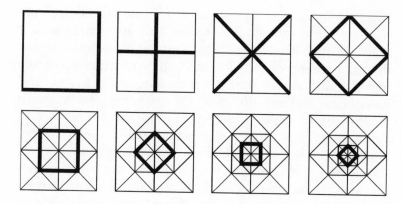

Decomposing a square.

through the bottom and top of the square. Use this distance to walk the compass around the square and find the middles of the four sides. Draw lines connecting these opposite midpoints to quarter the square. Also draw the square's diagonals. Diagonals are the means by which a square may be extended beyond itself or into its own depths.

The use of + and × between the midpoints and corners allows us to "harmonically decompose" or "diminish" the square in a quaternary rhythm, giving us lines and crossings that maintain the "root-two" geometry of generation, as in a woven cloth. Connect the four midpoints to make a tilted square. Notice the points where the diagonals cross it, and connect them to make a smaller square within it. Now notice where the smaller square's sides cross the +, and draw lines connecting these four points. Continue "weaving" this pattern, the ultimate symbol of material manifestation, in as small a scale as the point of your pencil will allow. Each square covers half the area of the previous, larger square.

Try doodling and shading this design to see patterns emerge. Extend lines to see greater geometric detail. The pattern is a powerful mandala to gaze at. When its geometry has been properly applied by skilled artists who know the rules for relating the scene to the geometry, it can bestow a structure with a gripping effect on viewers. Traditionally, only the images of deities align perfectly along the lines.

Part of a Babylonian clay tablet shows this construction as taught in the second millennium B.C.

"Worldly" characters are slightly "off" the lines to indicate that the world's material forms are only approximations of ideal, eternal archetypes. This placement gives the images of deities a feeling of serenity and "otherworldliness," while forms placed away from lines have a feeling of motion and life, of time and transitoriness. This was once part of a well-known design canon based on nature and the science of visual perception. It is periodically "rediscovered," used, and shunned throughout history. At its best, geometry gives the artist great freedom. But over time its use becomes mechanical and the art loses its spontaneity. Then artists shun geometry in favor of "free-form," each age finding its own expressions for the archetypes that motivate it.

Discover the "Goddess's Weave" in Natural Forms

A square with crossings has been constructed around both of these natural forms. One has been "harmonically decomposed." Do the same with the other to see how nature's designs express this geometry in their details.

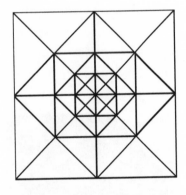

Harmonically decomposed square.

Tungsten atoms magnified two million times resemble watery ripples in a square pattern of energy vibration.

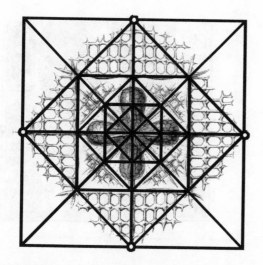

The skeletal design of a microscopic radiolarian follows the "lines of force" within a decomposed square.

Discover the "Goddess's Weave" in Goddess Art

For over fifty centuries, until recently, artists, craftspeople, and architects have consciously used the language of geometric symbolism as an underlying framework in their creations. They used different shapes for different themes. The decomposed square was often used to depict a goddess in her fertile, birthing, or nourishing character. Many examples from worldwide cultures use this same geometric symbolism, which arises spontaneously from everyone's own archetypal depths. Gods and goddesses were personifications of natural processes and principles at work in the cosmos and ourselves.

Each of the following figures is enclosed by a square with crossing lines within it. Some have already been analyzed. Harmonically decompose the square around the others to see how each was designed. As you do the geometry and see the grid crystallize, you'll see its relationship with the scene unfold and mesh. Use what you learn to design your own compositions.

*I*nspiration, even passion, is indeed necessary for creative art, but the knowledge of the Science of Space, of the Theory of Proportions, far from narrowing the creative power of the artist, opens for him an infinite variety of choices within the realm of symphonic composition.

—Matila Ghyka (Romanian royalty, educator, and writer)

*T*he aim of art is to represent not the outward appearance of things, but their inward significance.

—Aristotle

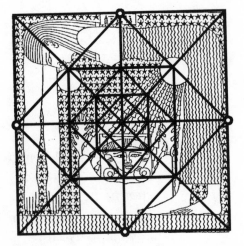

Arching with misplaced arms over the waters of chaos from which the cosmic order emerged, the Egyptian sky goddess and mother of the gods, Nut, swallows the sun at dusk amid the stars. She gives birth to it as dawn where its rays shine upon the face of the nurturing goddess Hathor on the horizon. (Painted ceiling relief, Temple of Hathor, Dendera, Egypt.)

Egyptian mother goddess Aset (known in Greek and English as "Isis") holds the glyph for "eternity" while kneeling upon the glyphs for gold and perfection. The scene's geometric center is just in front of her womb. (Sarcophagus of Tuthmosis III, c. 1450 B.C.)

In myth, the Egyptian vulture goddess Mut ("mother") was a virgin fertilized by the north wind. (Pendant from the tomb of Tut-Ankh-Amon.)

Sumerian pottery of the 5th
millennium B.C. depicts four
goddesses.

*The beautiful is a
manifestation of the secret
laws of nature . . . When
nature begins to reveal her
open secret to a person, he
feels an irresistible longing
for her most worthy
interpreter, art.*

—Johann Wolfgang Goethe
(1749–1832, German poet)

Lady of the Beasts. (Detail of
plate, Rhodes.)

*[Painters should] fix
their eyes on perfect truth
as a perpetual standard of
reference, to be
contemplated with the
minutest care, before they
proceed to deal with
earthly canons about
things beautiful.*

—Plato

Mater Makes Matter

. . . and the world was void and without form and darkness was
on the face of the deep.

—*Bible (Genesis)*

*If you would understand
the invisible, look carefully
at the visible.*

—Talmud

*Numbers are the sources
of form and energy in the
world.*

—Theon of Smyrna (Second
century A.D., Platonist
mathematician)

*The distribution of energy
follows definite paths
which may be studied by
means of geometric
construction.*

—Samuel Colman

Everything created shares the urge to create. Through the
miracle of birth the great cosmic creating process expresses
its archetypes and becomes more conscious of itself. By
bringing into expression the vision locked in our hearts we
resemble that process and grow in self-awareness. Through
geometric replication of nature's formations we can learn
about the archetypal principles inherent in universal design
and in ourselves.

In ancient mathematical philosophy, the first stirring of
the Monad results in the polar Dyad. Its geometric metaphor
is the construction of the *vesica piscis*. In myth, it is represented as the meeting of primeval parents.

All creation is due to polarity. Every birth occurs
through this interpenetration of positive and negative, light
and dark, male and female, god and goddess, electricity and
magnetism, and the two circles of the *vesica piscis* where geometric forms configure.

While modern science understands the universe as the
synchronization of electronic power with magnetic fields,
vibratory patterns of electromagnetic frequencies, and
energy bonds, the ancient bards sang of the union of the
Great Father and Great Mother. They saw the bright lightning of the sky galvanize the dark caverns of the earth into
activity, namely, fertilization of the land with nitrogen. Temples, mounds, and cairns around the world served as technological devices fertilizing the communities in which they
were built. The ancients personified their vision of the subtle forces of nature in myth. Where modern science observes
the triad of light, energy, and mass as $E=mc^2$, the ancients
mythologized a mystic marriage and birth in three stages: a
field or womb of light arises, it swirls as an energy pattern,
and physical forms precipitate upon the pattern. The
geometer replicates this cosmic configurating process with
three tools. When the precise union of circles occurs, the
three-stage process of birth begins, culminating in forms
clothed by the four states of matter.

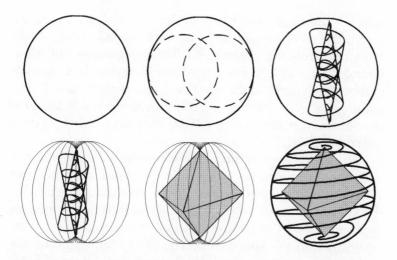

Geometric manifestation from light through energy to matter. A sphere representing undifferentiated light stirs, polarizing into opposites symbolized by two intersecting circles. The two poles swirl in opposite directions in mathematically precise rhythmic energy patterns, represented by spiraling cones. The sphere becomes a whirling energy field displaying lines of force upon which atoms configure to manifest material, visible volumes.

The womb of the World Mother stirs. Electronic power galvanizes magnetic energy into a state of activity, but there are no distinguishing features. The geometer uses the compass to draw the womb as a circle around the *vesica piscis* from its center. In the past it has been referred to as the Monad, Orphic Egg, Pneumatic Ovum, and Individualized Field.

Archetypal patterns arise. The electromagnetic field spins like a sun on its axis. Massless light whirls in mathematically precise rhythms and patterns. This is Plato's "world of pure knowledge, the realm of ideas" and the "world of wonder" described by seers. The Egyptians personified this field as the work of the architect/builder god Ptah. The Greeks called it Archeus, the builder of "first principles" (architect, arch, and archetype all derive from the

Orphic Egg surrounded by a spiraling snake, symbols of the stirring Monad, the womb, or field of light giving birth to our divine Self.

The worlds originate so that truth may come and dwell therein.

—Buddha

Geometry existed before the creation.

—Plato

What nature creates has eternity in it.

—Isaac Bashevis Singer (1904–1991, American author)

The right thing is to proceed from second dimension to third, which brings us, I suppose, to cubes and other three-dimensional figures.

—Plato

Greek word *arche*) according to the will of Zeus. In Hindu myth, the creative will of Indra was carried out by Vishvakarma, architect of the gods. The Freemasons call it the Grand Architect and Builder of the Universe. In Christian symbolism it is called the Mystical Light of the Logos. To the geometer's eyes, the *vesica piscis* inherently holds the proportions and patterns of the basic shapes even before they are constructed.

The goddess gives birth. The geometer makes visible the archetypal proportions inherent in the *vesica piscis*, helping them through the birth portal. In the electromagnetic field, the spiraling, massless photons thicken into tiny opposite-spinning whorls of protons and electrons that braid with neutrons into massive atoms. The precipitating atoms are tied in Borromean knots along the archetypal lines of force like threads in a carpet or beads on a loom, like morning dew on previously unseen spider webs, and like iron powder galvanizing near an invisible magnetic field. Atoms, the basis for the elements, precipitate in precise geometric patterns, like lace embroidery or Islamic tiles, along these lines of force. The geometries of nature's forms reveal the pattern of their energy fields. Natural structures, then, are the archetypal matrix made visible. As ideal geometric patterns *materialize* they're clothed with four thicknesses of garment, four densities of "mother substance," unfolding as light and thickening through the phases of gas, liquid, and solid. In mythic terms, the Demiurge is said to conquer chaos by dividing it into the four elements.

We can perceive the invisible lines of force by looking at the geometry of nature's forms. For example, dissolve as much salt into a bowl of hot water as it will allow and let it evaporate slowly over a period of days. A cubic lattice of salt crystals will precipitate in this matrix fluid. The more slowly the crystals grow, the larger they will be. Rapid evaporation produces small crystals as seen in Hawaii's volcanic obsidian stone, where molten lava hits water and cools immediately. The crystals are a picture in our consciousness made by our slowly perceiving nervous system, which we identify or project onto the geometric energy-configuration.

Nature's patterns are based on the mathematics of three-dimensional space. In essence, nature's creating process yields its fruit by giving birth to volume. Constructing the universe, therefore, involves the process of volumization.

The archetypal field where the patterns first appear is a sphere, a Monad. The geometry that arises within this sphere is obliged to manifest the Monad's principle of equality in all directions. Nature adheres to this principle by configurating primary volumes that divide the sphere equally in all directions. An experiment with a rubber ball and a piece of chalk will show that there are very few ways to put points around a sphere so that when they are connected by lines they form identically shaped surfaces, edge lengths, and angles. Nature's first expressions in three dimensions are such forms that fit perfectly within the sphere and present us with an identical view in all directions, no matter how you turn it. These forms, or "volumes," are based on surfaces with the same shape, either square, triangle, or pentagon. Ruled by the Monad, three-dimensional nature is structurally disposed to the geometry of equality in all directions.

There are only five volumes that fulfill this requirement of equality by repeating the identical corner angles, edge lengths, and surface shapes around a sphere. Although scores of stone models of each of the five types dating from 1500 B.C. have been found in Scotland, no one has ascertained their purpose. The five volumes were described by Plato in *Timaeus* and so are known to mathematicians as the "Platonic solids" and more formally the "regular polyhedra" ("many bases"). Four were identified by Plato with the four ancient elements, or states of matter, and the fifth, the quintessence ("fifth being"), represents their all-encompassing "cosmos." Their names derive from the number of faces they have.

"Heaven above."

The four "elements" of nature below.

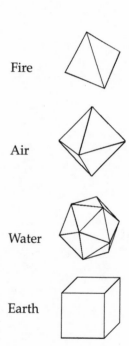

Fire

Air

Water

Earth

PLATO	VOLUME	SHAPE OF FACE	FACES	EDGES	CORNERS
"Heaven"	Dodecahedron	Pentagon	12	30	20
Fire	Tetrahedron	Triangle	4	6	4
Air	Octahedron	Triangle	8	12	6
Water	Icosahedron	Triangle	20	30	12
Earth	Hexahedron (cube)	Square	6	12	8

Mathematicians have long known there can be only five possible "equal divisions" of three-dimensional space. Highly respected, discussed, and wondered at in ancient times, the construction and study of these five volumes were considered the ultimate goal toward which the elementary constructions built, the culmination and pinnacle of ancient geometric and esoteric knowledge. Their construction comprised the final books of Euclid's *Elements,* a compendium that assembled and arranged the totality of Greek geometry without the original esoteric symbolism of the mathematical philosophers. His book began with the point, line, and plane and ended thirteen chapters later with the construction of these five volumes.

The first thing one might notice about these forms is that their faces are either triangles, squares, or pentagons. No other flat shape used alone will enclose a volume without gaps. These are the most economical expressions of space from which nature's designs derive their characteristics.

Variations on these five volumes in nature are virtually endless. They're the basis for all crystals, the orderly repeated arrangement of atoms.

The first manifestations of the universe are geometric. From hydrogen to uranium all ninety-two natural atoms of the periodic table that compose minerals and crystals are geometric. At the borderline between nonliving forms and living creatures are the cold, herpes, and AIDS viruses, which are not alive but act as if they were and also take form

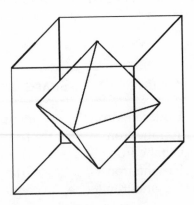

as Platonic volumes. Many microscopic life-forms share the geometry of these basic volumes. Each form's geometric expression is perfectly suited to its needs and purpose.

Recently, a golf-ball manufacturer discovered that by arranging the ball's "dimples" into a dodecahedral pattern, he could avoid having a seam around the ball's equator that slows and rotates it; thus, it can be made more aerodynamically stable and more accurate.

The dance teacher Rudolf Laban proposed a grammar of body movements based on the relationship between the proportions of the human body and the five Platonic volumes. He showed, for example, that the maximum angles through which human limbs move are identical to those angles of the twenty-sided icosahedron, the shape that gives the most symmetrical distribution of points, edges, and surfaces on the sphere. Thus Laban created dances whose architecture in time was identical with the spatial harmony of crystals. If a dancer had small lights on her limbs and danced in a dark room, the different Platonic volumes could be traced by her movements.

Many books have been written about these fascinating volumes. Particularly interesting are the relationships between them. For example, the corners of an octahedron fit in a cube and touch the center of each square face. Similar "duals" occur where the number of corners of one volume equals the number of faces of another. As this illustration shows, an octahedral crystal can gradually grow into a cubic shape, explaining the endless variety of crystals.

CONSTRUCT THE PLATONIC VOLUMES

To read about the Platonic volumes is not enough. Constructing them is simple. In building these solids the geometer replicates the precise mathematical proportions of the archetypal energy rhythms that nature configures as the various material forms. Three-dimensional space can be divided equally in all directions in only five ways by repeating one corner angle, edge, and surface shape. The cube, representing earth, is the only volume built entirely of squares. Three volumes have triangular faces and are known as *deltahedra*

Platonic volumes are related to each other. In this case an octahedron transforms into a cube.

("triangle-faced"). The dodecahedron has pentagonal faces and will be discussed and constructed in chapter five.

After constructing the models, hold and turn them in your hands. Get to know each volume by feel without looking.

*G*od ever geometrizes.

—Plato

The Birth of the Cube

The cube, or "hexahedron," is the best known of the Platonic volumes, the shape we know as six-sided dice but also found as salt and other crystals, as well as microscopic lifeforms. Geometers usually construct the cube either by folding paper or creating individual faces. Both produce fascinating results.

Salt crystal. Sodium chloride (NaCl) atoms tightly pack along cubic lines of force.

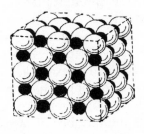

Paper-folding method: Construct six circles, intersecting *vesicas pisces*, as shown. Draw straight lines through the circles' centers (see illustration), extending some lines to reach the circumferences. Draw tab flaps where indicated. Cut along the lines and around tabs. Crease and fold the lines and tabs to make a three-dimensional cube. Glue or tape the tabs to the sides they encounter.

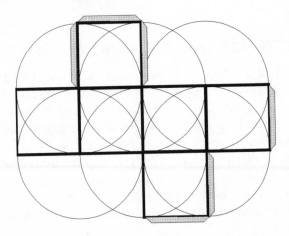

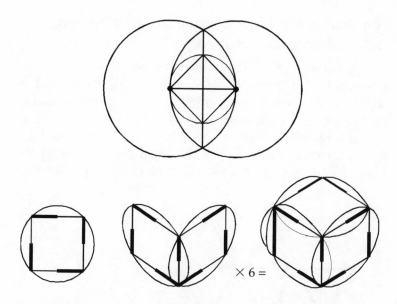

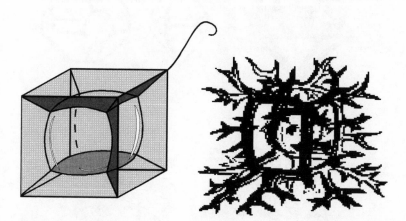

Face method: Bring a square through a *vesica piscis*. Cut out the circle around the square, and use it as a template for making six. Cut a slit along *half of each side* of each square. Slide the circles together through the slits to build the cube. You may wish to glue the rounded edges together to make the cube firm. Notice how the cube formed by the lines is inscribed or encased within the curved seams of a sphere.

He is invited to great things who receives small things greatly.

—Flavius (490–583, Roman legal writer and politician)

Make a cubic soap bubble. Bend a wire into a cube with a handle. Dip it into soap solution and let nature herself show you the minimal lines of force upon which she builds forms. If you capture a single bubble within it, a curved inner cube configures.

The Birth of the Three Deltahedra: Tetrahedron, Octahedron, and Icosahedron

Deltahedra refers to the three Platonic volumes whose faces are equilateral triangles. The paper-folding method is shown first. Construct intersecting *vesicas pisces*, and connect points as shown to produce triangles equal to the number of faces (four, eight, or twenty). Cut around the tabs, fold, and glue. This models the ideal geometry that the Many, the microbes, crystals, and molecules, approximate.

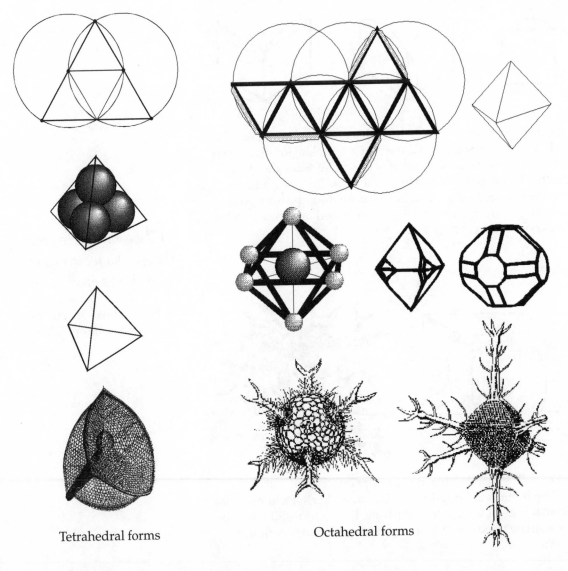

Tetrahedral forms   Octahedral forms

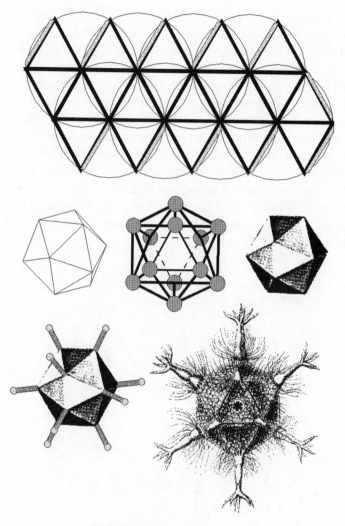

*T*ruth is so excellent, that
if it praises but small
things they become noble.

—Leonardo da Vinci

Icosahedral forms

Viruses associated with the
common cold, herpes,
measles, HIV take the forms
of Platonic volumes.

I think that modern
physics has definitely
decided in favor of Plato.
In fact the smallest units of
matter are not physical
objects in the ordinary
sense; they are forms, ideas
which can be expressed
unambiguously only in
mathematical language.

—Werner Heisenberg
(1901–1976, German
physicist)

The next method involves building solely by repeating the construction of a triangle inscribed within a circle. To make it, first construct a large triangle within the circles that create a *vesica piscis*. Find the center of this triangle by connecting each of its corners with the midpoint of its opposite side. Put the compass at the central point where they cross, and open the pencil to one corner of the triangle. Swing a circle around the triangle. Cut around the circle. Create one as a template to trace. Cut out four to construct a tetrahedron, eight for an octahedron, and twenty to build an icosahedron. Cut a slit along *half* of each side of each triangle as shown. Link them to configure the three volumes. Glue the curved flanges together. Notice how the volumes seem inscribed in spheres.

[T he Greeks] found in the construction of the five Platonic bodies in one sphere, the picture of the union of opposites, the work of the Demiurge, the creative artist of the world.

—Frederick Macody Lund

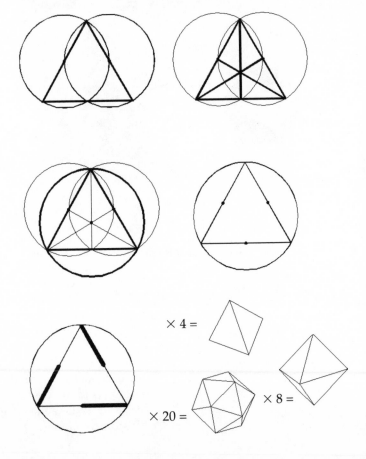

After building them, hold them, turn them, and look at them. They offer both an intuitive and intellectual education in harmony. First, get to know them by touch without looking at them. Then, for intellectual interest, you can verify the discovery of eighteenth-century Swiss mathematician Leonhard Euler that in each of these five volumes, *the number of corners plus the number of faces is always two more than the number of edges.* The numbers involved in these volumes have been found to be identical to the spatial representations of musical harmonies. Crystals, then, may be considered frozen music holding the proportions of musical intervals in the relationships of their corners, edges, and faces. The forms of nature configure as visual music. "Crystalloidal tectonics" is a term used to describe nature's rhythmic architectural designs in three dimensions.

The simple beauty of a crystal is, in part, its strict mathematical order. Through the study of number and shape we begin to apprehend cosmic order. Now we can see how the world virgin is fertilized, swirls into archetypal patterns, gives birth as the world mother, and then nourishes and clothes the geometric patterns with substance, the four ancient "elements," manifesting as atoms in configured patterns we can sense.

*T*he universe is the externalization of the soul.

—Ralph Waldo Emerson

FOUR CORNERS OF THE YOUNIVERSE

A common theme in worldwide mythology is the human as microcosm, a miniature model of the whole universe. Now dismissed as primitive, this view was long considered a key to the understanding of the Self. The ancient philosophers saw their inner lives arranged according to nature's own harmonies, for in nature, geometry, and their spiritual lives they discovered the same principles.

A fundamental map of ourselves is found in the mathematical intimacy between the Triad and Tetrad. The ancient mathematical philosophers saw themselves wherever three and four mingle. Three Borromean Rings intersect at four secondary spaces, just as light's three primary colors blend to produce four more. This simultaneity of three and four is a simple blueprint of ourselves: the triangle denoting a

*N*umber . . . should not be understood solely as a construction of consciousness, but also as an archetype and thus as a constituent of nature both without and within.

—Marie-Louise von Franz

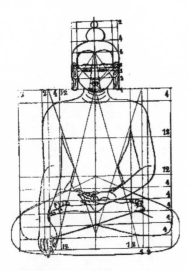

Tibetan proportions for depicting the Buddha crown triangle have been added.

divine trinity "above," or deep within, and below it a square unfolding the four elements of familiar human nature and body outward.

Trinity and quaternity is represented by the symbol of the pyramid on the Great Seal of the United States, whose divine eye in the triangle floats in light above its four-faced, square-based brick foundation. This "three above and four below" is also symbolized in the Sphinx's trapezoidal head-dress, which implies an unseen trinity above it (see chapter three). Our one Self has an inward and outward expression.

In painting and sculpture this theme was often depicted by delineating the head in a triangle and the body below within a square frame, as in this Tibetan plan of the proportions for correctly drawing the Buddha. (Geometrically decompose the large square to see its inner design.)

Note also the famous symbol of the Freemasons, the compass over the carpenter's rule. The compass is open sixty degrees, the angle of an equilateral triangle, above the rule's ninety degrees. The compass and square rule together symbolize our inseparable divine and human natures.

Few would suspect anything special about the design of doors, windows, and roofs. So common are weight-supporting triangular roofs and lintels over rectangular facades, doors, windows, and porticoes that we hardly notice them. But this is just the effect the enlightened archi-

The tetrad comprehends the principle of the soul as well as that of corporeality; for they say that a living creature is ensouled in the same way that the whole universe is arranged according to harmony.

—Iamblichus

A triangle above a square in art and architecture traditionally represents our deepest divine nature "over" the four "elements" of our human nature.

tects of ancient times wanted for embedding esoteric instruction about the self. The architecture itself, its proportion and ornamentation, held the esoteric teaching in plain sight, a reminder of their divinity to those initiated in the timeless wisdom that speaks of self-knowledge and spiritual transformation.

Copies of ancient architecture are scattered through our cities. The wisdom-teaching is still evident to those who notice and can read it. But this century's trend in urban architecture has been simply to hang walls on skyscraping steel frames. Triangular lintels are no longer used to support weight and pass along the traditional teachings that, until recently, had endured for thousands of years. By allowing the square and cube to eclipse the divine triangle from common sight, we've replaced a valuable body of knowledge with soulless architecture.

THE FOUR ELEMENTS WITHIN

The ancient philosophers saw the natural world of earth, water, air, and fire within us as the quatrain of our soul. In this inner symbolism, the four elements are seen as four levels of motivation within ourselves, a human tetrachord upon which the forces of nature play. The ancients symbolized it many ways (see chart below) in architecture, religion, mythology, and arrangements of society. Since the whole earth was symbolized by the human microcosm, it is *we* who are referred to as the "four corners of the globe" containing the "seven seas" (chakras), as well as the four stages of a spiritual journey and four levels of a temple or society. Adam's dominion over the earth (in Hebrew, *adama* means "earth") refers to our own mastery over the four realms of our *inner* nature, represented by beasts of the earth, fish of the seas, birds of the air, and "creeping things" (snakes and salamanders, traditional symbols of fire). To understand this symbolism each of us must become familiar with the ecology of the world within ourselves.

What is represented within us by the element "earth"? We form identities from pictures of the world as they appear to our senses, as surfaces and solid bodies. The earth nature within us relates to what is most dense, most slow to change, transform.

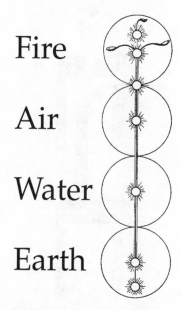

Fire

Air

Water

Earth

Seven "chakras," or centers of motivation, along our spine in four levels: earth (pelvis), water (belly), air (chest), and fire (head).

*T*he weather and my
mood have a little
connection. I have my
foggy and my fine days
within me.

—Blaise Pascal (1623–1662,
French mathematician and
philosopher)

The Hebrew letters YHWH
(English: "Jehovah") when
written vertically depict an
upright person in four levels,
image of the cosmos.

What "water" is within? The ancients saw our emotions in the "moods" of water: surging, rising and falling, turbulent, stormy, calm, sometimes laughing along a mountain stream or raining in tearful torrents, our emotions as part of our inner hydraulic cycle. Like water, emotions take the form of their container. Our emotions are shaped by our worldview, family traditions, opinions—in essence by our thoughts. These thoughts form the container that shapes our emotions. If your emotions are in a particular habitual pattern, find the thoughts and beliefs that have created the rigid pattern. Work to develop new thought patterns, and it won't take long to see the emotions take different forms.

The "atmosphere" within us is where our thoughts fly. Soaring flights of fancy, daydream castles in the air, the rarefied intellect, and clouded thoughts all refer to the influence of reason. The Egyptian artists symbolized this influence by depicting birds, sometimes netted, sometimes free and soaring upward. Our thoughts, like the wind, are a pollinating force, spreading ideas over our inner landscape.

The element of "fire" is represented within us by our intuitions. "A flash of insight," "it hit me like a bolt of lightning," and "seeing the light" are all phrases referring to a glimpse of our deeper Self, the very power by which we are conscious. Intuitions are most ephemeral, the most electric, the least dense, and the most difficult of the elements to discern. When we are most in touch with our deeper Self we are not forming pictures, thinking, or emoting but are simply aware of our inner "fire" and sense of purpose, experiencing a deep knowing beyond thought or emotion or sense. Fire is a symbol of purpose.

The various ways these four states of matter have been symbolized over the millennia as states within us are too numerous to explore in detail. In general, they describe a fourfold structure of the world within, symbolized by the *tetragrammaton* (Greek for "four letters"), the Hebrew letters YHWH (in English, "Jehovah").

These correspond with the four castes of Hindu Vedic society (priest, warrior, merchant, and servant) that comprise the body of society in the image of Purusha, the Cosmic Man. Similar in design, the structure of medieval European society separated monk, knight, burgher, and peasant,

Plantonic Volumes (Regular Polyhedrs)	Ancient "Elements"	Egyptian Animal Symbols	Egyptian Stone Symbolism	Traveling
Tetrahedron	Fire	Cat/Lion	Granite (Igneous)	Master
Octahedron	Air	Hawk	Alabaster	Driver
Icosahedron	Water	Crocodile	Limestone	Horses
Hexahedron (Cube)	Earth	Beetle	Sandstone	Carriage

Genesis: Adam's Dominion	Biblical Symbols	Journey of the Israelites	Transformation	Somatic Divisions
Creeping Things	Pillars of Fire	Promised Land	Intention	Head
Birds of the Air	Whirlwinds	Wilderness	Attention	Heart
Fishes of the Sea	Water	Red Sea	Concentration	Solar Plexus
Beasts of the Field	Dust	Bondage in Egypt	Meditation	Genital

Four Ages	Ancient Temple Structure	Rivers of Paradise	Life of Jusus	Human Body Systems
Gold	Holy of Holies	Honey	Resurrection	Nervous
Silver	Inner Chamber	Milk	Crucifixion	Respiratory
Bronze	Inner Court	Wine	Persecution	Circulatory
Iron	Outer Court	Water	Service	Digestive

Medieval Society	Hindu Castes (Body of the Universe)	Medieval Nature Spirits	What Do You Cling To? (Plato)	What Do You Desire? (Buddhist)
Monk (Clergy)	(Priest) Brahmin (Head)	Salamanders	Wisdom	Liberation
Knight (Military)	(King/Military)	Sylphs	Knowledge	Merits
Burgher (Commercial)	Kshatriya (Arms)	Undines	Opinions	Riches
Peasant (Agricultural)	(Merchant) Vaisya (Belly)	Gnomes	Ignorance	Pleasures
	(Laborers) Shudra (Feet)			

Many quaternaries symbolize the world as four elements, or levels. They represent four levels or centers of gravity within us with which we identify and express ourselves in the world. The purification of each level represents four stages of transformation and transcendence taught in myth and religion.

roles associated with the four humors and the four somatic divisions. Today, we might consider the four blood types.

Like the temple, mandala, and cathedral, ancient society was structured as a model of the cosmos and man. The Egyptians, for example, saw the scarab beetle, like most other insects, undergo four stages of transformation from silent egg to voraciously hungry larva to cocooned pupa to winged adult. They used this process in their hieroglyphics and mythology to symbolize spiritual transformation from a mundane life to initiation and the mummy-cocoon from which the priest-initiate emerged as a spiritually awakened adult.

As is evident in the stone symbolism in their architecture, the Egyptians were well aware of the four elements. In their modeling of the cosmos, "earth" was represented by sandstone, "water" by limestone (formed of compressed seashells), "air" by translucent alabaster, and "fire" in their stone pottery, statues, monuments, and temples by any igneous rock such as granite, basalt, and diorite.

The ancient fourfold symbolism appears in many forms from architecture to the origin of playing-card suits. It was intended to remind and teach the initiate about himself so he might, to the degree of his understanding, come to know himself as a model of the world.

Carl Jung recognized the fourfold nature of Hindu and Buddhist mandalas as symbolic of the Self and its journey through the four elements, which he labeled the "four orienting functions": sense, emotion, thought, and intuition. As part of their analysis, he had his patients construct personal mandalas. The first three elements are easily available to ordinary consciousness. The fourth, our intuitive fire, is like the star followed by the Three Wise Men. When our three lower elements become purified of destructive habitual patterns, they become "wise" and follow the spiritual light above and within.

The geometer who takes this study from its outer symbolic form to the inner sacred experience does so through meditation. Attention is directed not to any geometric construction but inward to one's own ecology. Self-discovery and self-knowledge require nothing outside us, only sustained attention directed to our inner landscape. It's not easy to watch sensations, emotions, or thoughts. We slip between them and mix them, tossed about by winds and waves of the world we have cultivated within. Focus on each of four levels of the body—the pelvis, gut, chest, head—and come to know them as realms of earth, water, air, and fire. Avoid encouraging the formation of pictures, scenes, images. Just concentrate on their *feeling*. Get to know your world as the ancient philosophers did, as a representation of the cosmos.

In constructing the patterns of the universe the geometer is building himself.

CHAPTER FIVE

PENTAD

Regeneration

It is a frequent assertion of ours that the whole universe is manifestly completed and enclosed by the Decad, and seeded by the Monad, and it gains movement thanks to the Dyad and life thanks to the Pentad.

—Iamblichus

Where plants have five-fold patterns, a consideration of their souls is in place.

—Johannes Kepler

The mathematical rules of the universe are visible to men in the form of beauty.

—John Michell (1933– , English philosopher, antiquarian, geometer, writer)

. . . the irregular and vital beauty of the pentagram.

—Claude Bragdon (1866–1946, American architect, philosopher, and writer)

Let us build altars to the Beautiful Necessity, which secures that all is made of one piece.

—Ralph Waldo Emerson

There is a geometry of art as there is a geometry of life, and, as the Greeks had guessed, they happen to be the same.

—Matila Ghyka

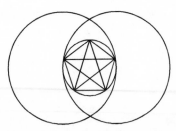

Birth of the Pentagon and
Pentagram Star.

COMING TO LIFE

We now arrive at the mathematical middle of our journey to ten to dwell with the principles symbolized by the number five, which the Greek philosophers called Pentad. Beyond the Monad's point, the Dyad's line, the Triad's surface and the Tetrad's three-dimensional volume, what remains? The Pentad represents a new level of cosmic design: the introduction of life itself.

We encounter the Pentad in all expressions of fiveness, from our hand to the stars on flags. Geometrically, the Pentad is born through the *vesica piscis* as the pentagon and pentagram (five-pointed) star, as whirling spirals, and as the three-dimensional *dodecahedron*, the fifth of the five Platonic volumes having twelve pentagonal faces. This last form has been known to philosophers as the *Quintessence* ("fifth being") of nature, encompassing and infusing the four elements—solids, liquids, gases, and electronic fire—with the life they cannot create by themselves alone. Pentagonal symmetry is the supreme symbol of life. Many living forms, plant, animal, and human, display the clear geometry of the Pentad in their structure. The four elements supply the materials of their configuration, but the Pentad carries the flag of life. We'll see how it links them with the mystery of the mathematical infinite, which is the mystery of life itself: the ability to regenerate.

On our way we'll pass through fascinating historical topics known as the Fibonacci numbers and the golden mean for insight into the secrets of harmony and beauty, the accord of the parts with the whole. Harmony and beauty are essential part phenomena.

Pentagonal symmetry has been long revered due to its profound insight into living nature and to the powerful psychological hold it has upon people throughout the world. It manifests itself in surprising ways and places in art, crafts, architecture, religion, magic ritual, national icons, and much else that is rooted deep within us.

Enter these pages with a feeling of wonder at seeing the cosmic creating process reach its pinnacle of expression.

The feeling of awe and sense of wonder arises from the recognition of the deep mystery that surrounds us everywhere, and this feeling deepens as our knowledge grows.

—Anagarika Govinda

The most beautiful thing we can experience is the mysterious. It is the source of all true art and science.

—Albert Einstein

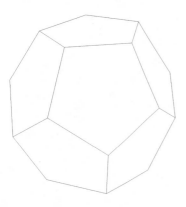

SEEING STARS

It's of little use, at first, to try to understand the principles of the Pentad by examining the ways it shows up in universal sayings and mythology. Can we immediately grasp why the Chinese revere the "five bats of happiness" or why David armed himself with five stones against Goliath? It's better to consider where we see it graphically as its geometric icon: the five-pointed star.

Think about where you see stars.

Perhaps in school your teacher put a gold star on work well done. Worldwide, the star recurs as a symbol of excellence in rating systems. We speak of five-star hotels, films, plays, and restaurants. Actors and performers who excel achieve "stardom" and are referred to as "movie stars," "stars of stage and screen" giving "stellar" performances. Critics are always looking for "bright new stars" rising on the horizon. The imagery is not limited to drama: "sports stars" and "rock stars" are acclaimed by their "starstruck" fans. "Superstar" status is conferred on those especially esteemed. Unlike lunacy, which has a negative connotation, stardom conveys a deep-seated association with excellence and brilliance.

The five-pointed star wells up spontaneously as a worldwide symbol from the deep archetypal level common to everyone. Have you ever wondered why the flags of over

sixty nations include a five-pointed star? Flags affect people deeply. There must be something very powerful in the star icon if so many nations with conflicting ideologies each claim this same symbol as their own. One would expect each country to design its flag to be individual, unique. What is it they all vie for? Vexillologists tell us that the appearance of star-spangled flags continues a ritual dating back millennia, when flags were seen as magical symbols of power and invulnerability. Everyone wants to be associated with excellence; the star's appeal emerges in the ceremonious respect accorded flags and five-part phenomena.

We encounter the star pattern ceaselessly in commercial logos. Modern advertising agencies spend billions analyzing aesthetic preferences and have rediscovered the universal appeal of pentagonal symmetry. Something inside us clearly resonates with the fivefold form. Just look at the open embrace of a well-constructed pentagonal star and notice the feelings it evokes within you. The world recognizes and respects the five-pointed star for the same reasons: it imparts to everyone an irresistible impression of excellence and power.

From ancient times this star has been a magical symbol for warding off evil in Babylon, Egypt, Greece, India, China, Africa, the Americas, and elsewhere.

In literature, Goethe's Faust drew a pentagram star to ward off the devil, but as it was constructed imperfectly, Mephistopheles was able to approach him. The magical association of this star continues today in military matters as a ranking system ("five-star generals") among those, for example, who meet at the Pentagon, the world's largest office building, near Washington, D.C. Along with excellence and power the pentagonal star induces a feeling of authority, further explaining its military presence and appearance on flags and in commercial logos.

When the star is inverted it's called a *Drudenfuss* (German for "witch's foot") and appears in magical ritual and witchcraft. It has been associated with devils and demons since the rebel Egyptian god Set resisted spiritual aspiration to become the adversary of his brother Osiris and nephew Horus. Set had many attributes associated with the modern image of a devil: horns, coarse red hair, hooves, a tail. He

"You can trust your car to the man who wears the star . . ." (1950s Texaco Oil Company jingle)

and his hordes resist truth, order, justice, and righteousness. His forty-two attributes are actually symbols of the down-reaching elements of our own psyche that must be purified. The Egyptian "underworld," the *duat,* whose hieroglyph is a circle surrounded by a five-pointed star within a circle, is the mythical nighttime place where the "perishable" stars go when they sink below the horizon at dusk. The *duat* recalls our own spiritual sleep from which we can awaken. The Egyptian star glyph *without* the circle signified either a nighttime star, a door, or a teaching, an intriguing combination of ideas in one glyph. The sun's inevitable birth into dawn and daylight (symbolizing our own spiritual arising) is resisted in the *duat* by hideous adversaries (illusions within us). In ancient Greece, the star was associated with Pan ("All"), who also had horns, red hair, hooves, and a tail, representing the lustful fertility of nature, which induces *panic* and *pan*demonium ("all the demons") in prim people. Later in Western culture Pan's sexuality became the great temptation of the modern devil and the star was inverted to imply the reverse of, and resistance to, its positive qualities of excellence and goodness.

Clearly, the pentagonal star has been a powerful psychological icon in the past and present and undoubtedly will be so in the future. It conveys an impression of excellence, authority, and uprightness and strikes us all with a power and strength that can only come from the deep level of archetypes within us.

But exactly what is it about the five-pointed star, and not other geometric shapes, that appeals to us in this way? What is the source of this star's power and excellence? From where does the star derive its authority? What are the specific principles of the Pentad that touch us so deeply? What is the message of this star?

*T*o hold, as 'twere, the mirror up to nature.

—William Shakespeare

THE WORLD'S OLDEST LIVING STAR

Fivefold symmetry fills a special place in nature's design lexicon. The five-pointed star's association with excellence, power, and authority is manifested in nature as beautiful and efficiently designed forms. Pan's fertility was symbolized by the five-pointed star because the shape is a key to nature's fecundity and generative powers, providing the freshness, fullness, and fertility of life. The archetype of five-ness expresses itself as living forms and in the attributes of living creatures. It was for this reason that the pentagram star was a symbol of humanity and health to the Pythagoreans, who wrote the name of the goddess Hygeia around its points. Pentagonal symmetry is the flag of life. When it appears in nature we are seeing the archetype of the Pentad expressing life's excellence and authority.

We can't help but encounter stars and pentagonal structure throughout nature. The overall design of any leaf fits within a stretched or compressed pentagon. Leaves have the same symmetry as hands. Look at any bush or tree and visualize its leaves as open hands catching sunlight.

The Pentad is particularly comprehensive of the natural phenomena of the universe.

—Iamblichus

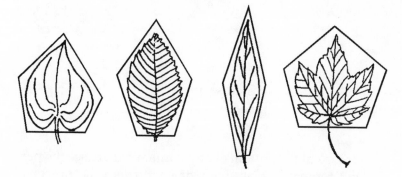

Leaves configure their archetype as stretched or compressed pentagons. The One underlies the Many.

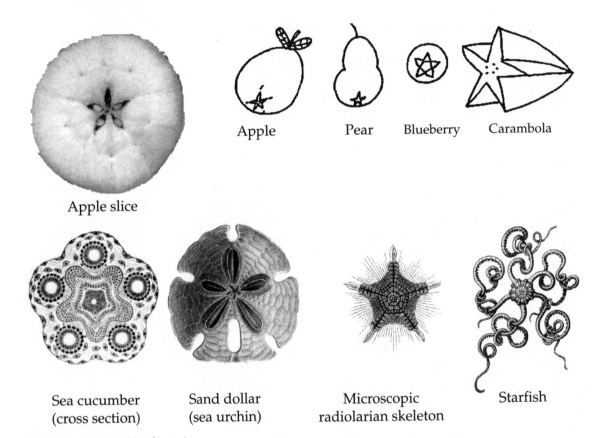

Apple slice

Apple   Pear   Blueberry   Carambola

Sea cucumber
(cross section)

Sand dollar
(sea urchin)

Microscopic
radiolarian skeleton

Starfish

Pentagons abound in the animal kingdom. The phylum Echinodermata, represented by the starfishes, sea cucumbers, and sand dollars shown here, is particularly obvious in its pentagonal symmetry.

Or look at the bottom of any familiar fruit to see the pentagonal remnant of its five-petaled flower. The flower of every edible fruit has five petals (though not all five-petaled flowers yield edible fruit). Before you eat an apple or pear, cut it in half at its equator to see the star pattern in which its seeds, the holders of life itself, are arranged.

When we look at any leaf, flower, or fruit we're seeing the invisible energy web of the archetype made visible as a pattern of living cells.

. . . thou cunning'st pattern of excelling nature.

—William Shakespeare

A STAR IS REBORN

We can gain insight into the Pentad's excellence by geometrically constructing its representation, the pentagon. In this shape, its most elementary visual form, we'll discover spe-

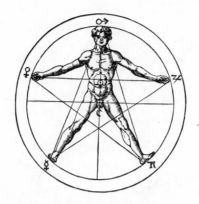

The Roman numeral representing five, symbolized by the letter V, derives from the shape of the space between the open thumb and fingers. The Roman numeral for ten, the letter X, is actually two V's.

The human body clearly expresses the Pentad's symmetry in its five senses and five extensions from the torso, each limb ending in five fingers or toes.

Pentadactylism, the fivefold symmetry of hands, paws, and claws, expresses the Pentad, life.

cial inner relationships that will help explain the power of the star and its archetype.

A pentagon is among the more complex constructions to be brought through the *vesica piscis*. But it needn't be difficult. Of the many ways to construct a pentagon, the simplest occurs every time we tie our shoelaces or a bow tie. The initial knot produces an internal "good-luck" pentagon. To see it, tie a strip of paper into an ordinary knot. Tighten it carefully and press it flat as you gently pull. Hold it up to the light to see its inner pentagram star. Tie five such knots at regular intervals along a long strip of paper to make a pentagonal frame with a star at each corner.

Another method is to start with a paper square representing the "four states of *mater*." Fold and cut it as shown in the illustration to produce stars in positive and negative space.

Construction using the geometer's three tools requires more complex steps than we experienced in constructing a triangle or square, as well as a deeper knowledge of the relationships within the *vesica piscis*. Few would discover it acci-

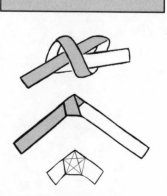

Tying a pentagonal knot.

Classic puzzle: How can you plant ten flowers in five rows and have exactly four flowers in each row?

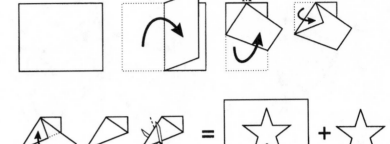

Answer: Plant them at the five corners and five crossings of a pentagram star.

dentally or by trial and error. The oldest construction on paper was taught by Euclid in his *Elements* (c. 300 B.C.), although much older Babylonian clay tablets describing the steps have been unearthed.

Periodically throughout history the construction of the pentagon has been so revered that it was kept secret by various societies who recognized its predominance in nature and its potent effect on human nature. They studied the Pentad's principles in geometry and nature, and its psychological effect, deeply enough to see in its application a knowledge that could be misused. Around 500 B.C. the pentagram star was the sign used by advanced members of the Pythagorean society to recognize one another, and one thousand years later it was a guarded teaching tool only taught orally, not written about, by the crafts guilds who infused its symbolism into the designs of Gothic cathedrals. It wasn't until 1509 when the monk Fra Luca Pacioli, inventor of double-entry bookkeeping and the mathematics teacher of Leonardo da Vinci, published *De Divina Proportione*, that the method of its construction and unique geometric properties was revealed to artists and philosophers in public.

It's out in the open again. Use the geometer's tools, and follow these steps (see illustrations on the next pages) to construct the pentagon and patterns that come from it.

Some constructions of the pentagon you may come across are actually only approximate, although very close, but this method produces one that is perfectly proportioned.

Construct the pentagon.

Follow these steps with the greatest precision and the sharpest pencil point.

1. Construct a large *vesica piscis.* Draw a line connecting the circles' centers and a vertical line up the middle.

2. Put the point of the compass where the lines cross, and open it to the center of one of the circles. Turn a circle within the *vesica piscis* as you did to construct a square. The pentagon represents the "fifth essence," life itself, encompassing and interpenetrating the four states of matter.

3. Using the same compass opening, put its point at the center of the large circle, and turn another small circle, making a small *vesica piscis.*

4. Draw a vertical line up the small *vesica piscis.* Find its center where it crosses the line between centers.

5. Put the unmoving leg of your compass at the center of the small *vesica piscis,* and open its pencil to the point atop the small circle within the large *vesica piscis.* Swing the compass downward until it crosses the horizontal diameter (or complete the arc to the full circle).

6. Put the compass point atop the small circle, and open the pencil to the point created on the circle's diameter. Swing the compass upward until it crosses the small circle. This new point and the point atop the circle are the pentagon's first two corners. We have organized Monad by crystallizing the Pentad within it.

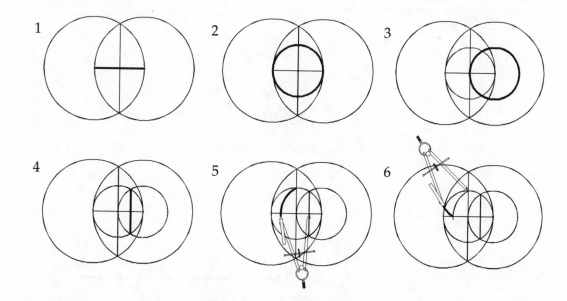

7. Using the same compass opening, walk the compass around the circle to find each new point until all five points are identified. Your pencil's final step should end precisely at the top where it began. (Helpful Tip: This construction requires great precision. If your com-pass was off even slightly at any stage, the difference will be magnified and the pentagon lop-sided. To double-check, walk the compass around the circle in the oppo-site direction. If there's any differ-ence between the marks made in each direction, the actual points of the penta-gon are found halfway between them.)

8. Connect the five points to make the pentagon visi-ble. Welcome the emissary of living structures, the geo-metric flag of life itself.

9. Connect alter-nating points to reveal the penta-gram star.

10. A variety of pentagonal forms can be constructed by this method, each of which expresses the Pen-tad's principles. Practice construct-ing them. Save one as a template for producing others quickly.

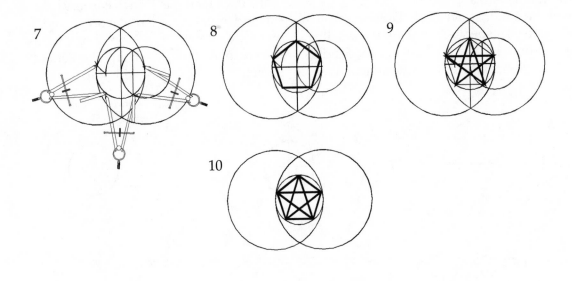

BUILD A DODECAHEDRON

The fifth Platonic volume is the *dodecahedron* (Greek for "two plus ten faces"). Like the four other volumes, it expresses the principle of equality in all directions, having twelve identical regular pentagonal faces, thirty equal edges, and twenty vertices meeting at three-corner joints. It is closest of the five forms to having the volume of the Monad's perfect sphere. As "quintessence" it isn't *above* the four elements, the configured states of matter, but encompasses them, infusing the force of life and excellence throughout their structures. Even nonliving crystalline minerals take part in the cycles of life when they're ingested by plants, animals, and humans, thereby becoming incorporated into our bodies. Like the dodecahedron, life's cycles encompass all the universe in its processes.

Today, we encounter dodecahedra, if at all, as those twelve-month desk-calendar paperweights and as special dice. The ancient mathematical philosophers revered the dodecahedron for its geometric properties and symbolism. Like a pomegranate ready to burst forth its seeds, the dodecahedron represents the archetype of life and fecundity made visible. It's a three-dimensional pentagonal web on which life expresses its fullness.

The best way to appreciate a dodecahedron is to build one. Unlike triangles, squares, and hexagons, pentagons cannot fill flat space without leaving gaps. But twelve pentagons will enclose three-dimensional space as the Quintessence encloses the elements.

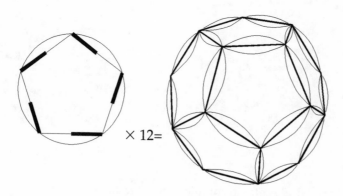

× 12=

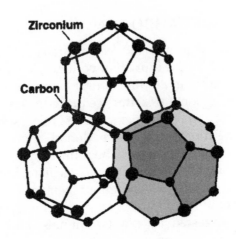

*S*ocrates said that, from
above, the Earth looks like
one of those twelve-patched
leathern balls.

—Plato

Scientists have recently constructed dodecahedral molecules
using carbon and various metal atoms that join to form larger
molecules. Recently, a third form of pure carbon beyond graphite
and diamond was discovered and formally named "Buckmin-
sterfullerenes," or "Buckyballs," after the mathematician, architect, inventor, and visionary R. Buckminster Fuller, who used
similar shapes for his geodesic domes. A Buckyball consists of
sixty carbon atoms, *five* around each corner of a dodecahedron.

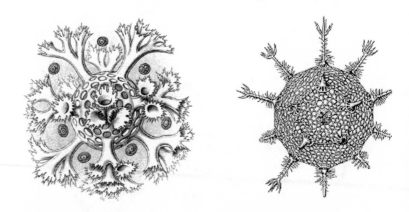

Dodecahedral skeletons of microscopic radiolaria.

Hold and turn the dodecahedron to appreciate the beauty of its lines and surfaces. Get quiet within yourself and be aware of feelings this fifth Platonic volume evokes.

WHOLES IN NATURE'S FABRIC

Avoid extremes—keep the Golden Mean.
—*Cleobulus of Lindus (sixth century B.C., one of the Seven Wise Men of ancient Greece)*

Nature expresses the cosmic philosophy through geometry. A deep look into the star's unique geometry will take us even closer to the core of the archetype of the Pentad. There, perhaps, we can discover the source of its power and the secret of its appearance in nature and its psychological effect on us.

A clue is found in the ability of vegetation and certain creatures to regenerate another whole like themselves from a part, not only from their seeds. For example, when put near soil, the leaves of some plants send out roots and grow whole new plants. Flatworms and other creatures with primitive nervous systems also regenerate new creatures from their parts. Most familiar is the ability of starfish to regrow a lost leg, while the lost leg grows a whole new body.

We can see the principle of regeneration, the whole in the part, in many places. The veins in a leaf reveal the branching pattern of the whole leafless tree. Each ridge of a fern's leaf models at once the whole leaf and the whole plant. Each seed on a dandelion's head mimics the whole head in miniature. Who hasn't noticed that the tips of broccoli and cauliflower resemble miniature trees?

Look at the top of a full broccoli or cauliflower head while squinting and you can see the overall pentagonal groupings. Five major clumps contain constellations of smaller and smaller pentagonal structures: each clump is made of five smaller clumps, and so on. This is the rhythm by which vegetation grows, a pulsing fivefold rhythm. Each successively smaller branch models the others and the whole.

This whole-in-the-part principle is the basis of bonsai, the Japanese art of growing miniature trees and bushes from the branches of a large tree. A tree's leaves swirl around their

The veins of a leaf, dandelion seeds, and branches of broccoli model the whole plant in each part.

twigs in the same pattern that twigs swirl around branches, as branches around the boughs, as boughs emerge from the trunk. Each taken by itself is a model of the whole. The next time you can dwell with a bush, look for this structural self-symmetry. Distinguish each level, and build your vision until you can see all the structure comprehended as the whole. Cognizance of this harmony in nature and mathematics attunes us to harmony at our own core.

Today, we limit the word "symmetry" to mean a simple mirror image, such as our two hands. But the ancient Greek word *symmetria* literally meant "alike measure," a term perfectly suited to these expressions of self-similarity on different scales.

This fascinating characteristic of living forms curiously corresponds with the geometric property of the pentagram to replicate forever smaller and larger versions of itself. The star replicates and inverts itself in its own central pentagon, which in turn has its central pentagon, in an endlessly diminishing series of inverting stars.

Flip the central pentagram outward and it transforms at each of the star's five points into a whole pentagram. These become nuclei for smaller stars to replicate within, in its center and around its points, multiplying endlessly. The Pentad holds the principles of the geometry of regeneration. It is

nature's way of producing endless variety using a single self-reproducing scheme.

Pentagons perpetuate their own image in endless detail. We are drawn into them as into the eyes of a lover, as we see the whole by looking at the part. Know one star and know the whole.

This self-similar accord, repeating the same shape on different scales, is the basis of fractal mathematics, which is at the heart of modern chaos theory. For example, the microscopic bumps on a grain of sand at the seashore resemble an aerial view of the whole beach's shoreline. Each small part of a turbulent system turns out to be a model of the whole and can be described mathematically by self-replicating formulas. Even in chaos we find the signature of cosmos.

This part-as-whole occurs in the scientific art of producing holograms. Unlike ordinary photographic negatives, in which a piece from the corner holds only a small part of the picture, any piece cut from a hologram contains the information of the whole scene in miniature.

Explore the Pentagon's Regeneration

Construct a pentagon or trace the five corners of any illustration, and connect them to enclose a pentagram star. Then do these activities:

      Doodle pentagons and pentagrams. Explore pentagons on your own. Draw them freehand, with each hand, as a continuous line. Extend lines and

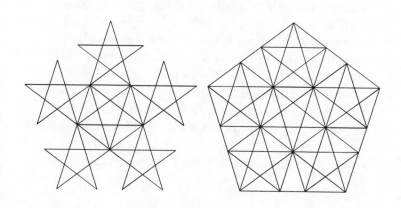

The pentagram exhibits self-similarity in endless variations. Explore its regenerating geometry on your own by subdividing each part into smaller pentagrams.

I believe the geometric proportion served the Creator as an idea when He introduced the continuous generation of similar objects from similar objects.

—Johannes Kepler

I ntuition is the clear conception of the whole at once.

—Johann Lavater (1741–1801, Swiss poet, mystic, and philosopher)

connect points to see how the star regenerates itself endlessly into smaller and larger models of the whole. The pentagon is extremely rich in endlessly fascinating self-replicating geometry that is accessible with a sharp pencil point. Discover its galaxies of stars, all identical in shape, differing only in size.

1. Decompose and fold a pentagon.

2. Geometrically construct a pentagon and connect its corners to inscribe a pentagram star.

3. Connect the corners of its central pentagon to draw a smaller but inverted pentagram.

4. Extend the lines of this inner pentagram outward so that they meet the sides of the large pentagon (see illustration on page 111). Use this as a template to make other pentagrams.

Use the straightedge to crease sharply all the lines both ways. Feel and experiment with the pentagon's tendency to fold various ways. Get all the points to meet at the top and let them spring back; this should remind you of a flower bud opening.

Cut the star into pieces. Construct another star, and cut the whole into pieces along the lines. Rearrange them to make smaller and larger stars and pentagons. Notice their interrelationships. Parts added together match exactly with larger ones.

Construct dodecahedra. Repeat the construction, decomposing a pentagon. Cut a slit along *one* side of each of the small outer triangles (see illustration on page 113). Fold the sides upward, and slip each small triangle under (or over) the piece next to it. Glue or tape it there as a tab, connecting the whole into a "basket" resembling an open flower. Two of these baskets joined face-to-face will form a dodecahedron. Or twelve of these baskets can be joined back-to-back by their sides to make a larger dodecahedron. With a needle and thread you can sew straight lines in three dimensions between

the points and corner folds of the pentagonal "basket" to reveal unexpected and beautiful geometric
relationships within it.

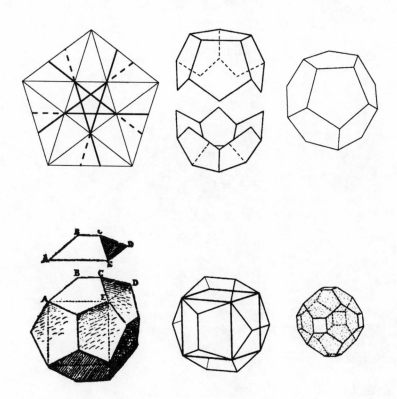

Discover Pentagonal Self-Symmetry in Natural Forms

Each life form below is enclosed within a pentagon. Use a
straightedge to connect its five corners to inscribe a pentagram star. Further decompose the pentagons and notice the
remarkable detail with which the creature expresses its
archetypal design, and how each small part is structured as
the whole. Some examples have been completed to different
degrees to show you how it's done.

114

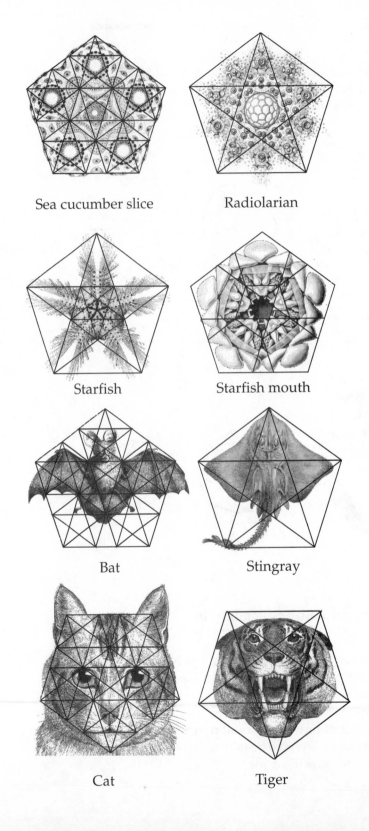

Sea cucumber slice

Radiolarian

Starfish

Starfish mouth

Bat

Stingray

Cat

Tiger

THE RABBIT RIDDLE

It is worthwhile for anyone to have behind him
a few generations of honest, hard-working ancestry.
        —*John Phillips Marquand (1893–1960, American novelist)*

The ways in which numbers behave and interact reveal the
most essential workings of their archetype. A look at the
actual numbers and processes involved in pentagonal self-
symmetry will take us as close as we can get to understanding the principles of the Pentad. The key to this regenerative
pentagonal geometry can be found in a particular self-
generating number series known as the Fibonacci (fib-oh-
NAH-chee) sequence.

Until the twelfth century, European commerce, banking,
and measurement relied upon the clumsy system of Roman
numerals, a vestige of the ancient Roman Empire, to calculate with. Try multiplying or even adding two Roman
numerals together to see how difficult it is even to begin. But
in the year 1202 a book, *Liber Abaci* (*The Book Of Computation*),
was published by a mathematician and merchant, Leonardo
of Pisa. This book convinced Europe to convert to the numerals we use today, zero through nine, known by their origin
and path to the West as the Hindu-Arabic numerals. Because
his father was nicknamed Bonaccio ("man of good cheer"),
Leonardo was known by the Latin for *son of Bonaccio,* "filius
Bonaccio," contracted to "Fibonacci."

Of interest to us from his book is an ancient number puzzle about breeding rabbits. Although there's no evidence
that Fibonacci suspected it, this puzzle holds the key to self-
replicating growth in geometry and nature.

The "rabbit riddle" essentially goes like this, The date is
January 1 and a pair of newborn rabbits, a male and female,
are in a pen. How many pairs of rabbits will there be one
year later if newborn rabbits take exactly one month to
mature, at which time they immediately mate, gestate for
another month, then give birth to another pair like themselves, and mate every month thereafter? Every newborn
pair repeats this pattern of monthly maturing, mating, ges-
tating, and breeding other identical pairs, which likewise do
the same. (We assume an ideal situation: no pair dies or
deviates from the pattern.) A visual sketch of their family
tree is the best way to start.

We soon become bogged down in the drawing of symbols representing rabbit pairs. But Fibonacci wanted us to notice the *numbers* representing their growth. A count of the newborn, mature and total rabbit pairs each month produces an interesting pattern, which recurs for each pair. When we notice its secret—that each two consecutive numbers add together to produce the next number—we can solve the puzzle easily by continuing this pattern to the next January 1.

This same pattern would recur if someone decided to spread a rumor in a crowd. If he thinks about it for thirty seconds and then tells it to someone who hasn't heard it every thirty seconds thereafter, and if everyone who hears it waits thirty seconds and then also passes it along every thirty seconds, the numbers of knowers, tellers, and hearers will increase by terms of the Fibonacci sequence. Other phe-

○ Newborn pair

● Mature pair

Rule 1
In one month each newborn pair becomes mature.

Rule 2
Each mature pair breeds a newborn pair and mates again.

Follow the two "rules" of growth to find how many pairs of rabbits there will be on January 1. Notice, in the fashion of self-similarity, that if you break off any "branch" it resembles the whole family tree.

		Total	Pairs
January	1	0	1
February	0	1	1
March	1	1	2
April	1	2	3
May	2	3	5
June	3	5	8
July	5	8	13
August	8	13	21

Answer: On January 1 there will be 233 pairs of rabbits.

nomena from handshakes in a crowded room to robot-building robots would do the same.

The Fibonacci sequence actually begins with two terms, zero and unity, nothing and everything, the Unknowable and the manifest Monad. These are the first two terms. Their sum, another unity, is the third term. To find each next term, just add the two latest terms together. This process produces the endless series 0, 1, 1, 2, 3, 5, 8, 13, 21, 34, 55, 89, 144, 233, 377, 610, . . .

Look at this series. At first glance we see a chain of numbers. But look beyond the visible numbers to the self-accumulating process by which they grow. The series grows by accruing terms that come from *within* itself, from its immediate past, taking nothing from outside the sequence for its growth. Each term may be traced back to its beginning as unity in the Monad, which itself arose from the incomprehensible mystery of zero.

This principle of ongoing growth-from-within is the essence of the Pentad's principle of regeneration and the pulsing rhythms of natural growth and dissolution. It appears in plants, music, seashells, spiral galaxies, the human body, and everything associated with the fiveness in nature.

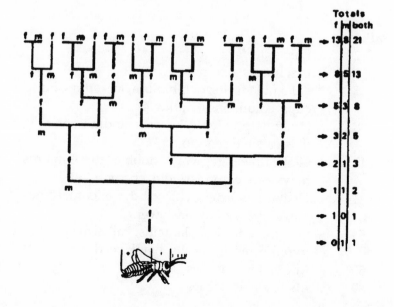

Genealogy of the drone (male) honeybee. Some insects and microscopic life experience *parthenogenesis* ("virgin birth") whereby unfertilized females give birth to males. But fertilized females always give birth to other females. This pattern maintains the dynamic balance of the hive. The family tree of any bee branches in the accumulative Fibonacci growth rhythm. And each branch resembles the whole family history.

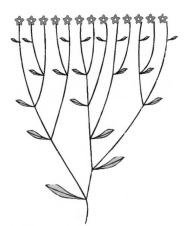

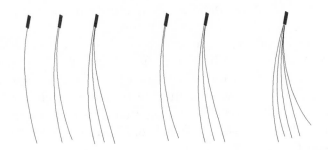

The average number of petals on each type of flower in a field will be Fibonacci numbers, as are the numbers of pine needles accumulating as clusters in different pine species.

Sneezewort (*Achillea ptarmica*). Each branch lengthens through time and then reproduces another like itself, which repeats the pattern of Fibonacci branching.

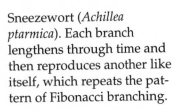

The piano keyboard is a metaphor of accumulating vibration and so is structured by terms of the Fibonacci sequence. The thirteen-note chromatic musical octave consists of eight white keys (whole tones) and five black keys (sharps and flats) arranged in groups of threes and twos making one full octave.

# Petals	Flower
2	Enchanter's nightshade
3	Iris, lilies, trillium
5	All edible fruits, delphinium, larkspurs, buttercups, columbines, milkwort
8	Other delphiniums, lesser celandine, some daisies, field senecio
13	Globe flower, ragwort, "double" delphiniums, mayweed, corn marigold, chamomile
21	Heleniums, asters, chicory, doronicum, some hawkbits, many wildflowers
34	Common daisies, plantains, gaillardias, pyrethrums, hawkbits, hawkweeds
55	Michaelmas daisies
89	Michaelmas daisies

THE GOLDEN MEAN

Write out the first few terms of the Fibonacci sequence. Draw a line under each to make it a fraction, and underneath write the Fibonacci sequence shifted back one term.

$$\frac{1}{0} \quad \frac{1}{1} \quad \frac{2}{1} \quad \frac{3}{2} \quad \frac{5}{3} \quad \frac{8}{5} \quad \frac{13}{8} \quad \frac{21}{13} \quad \frac{34}{21} \quad \frac{55}{34}$$

Use a calculator to divide and convert each fraction into decimal form.

A graph of the results shows each term getting closer to an ideal of 1.61803398875 . . . or rounded off to 1.618 or even 1.62. Notice how it begins crudely, pulsing far over then

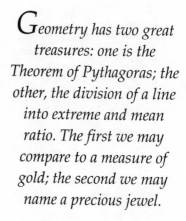

Geometry has two great treasures: one is the Theorem of Pythagoras; the other, the division of a line into extreme and mean ratio. The first we may compare to a measure of gold; the second we may name a precious jewel.

—Johannes Kepler

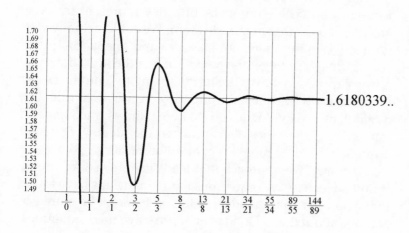

1.6180339..

The Fibonacci sequence has fascinating number properties. For example, its part resembles the whole. Use a calculator to divide the twelfth term, 89, into unity. The result is an endless decimal that, broken into parts, replicates the entire Fibonacci sequence: (1/89 = .011235813213455889144 . . .)

Historical names of the Golden Mean:

Plato	*the section*
Euclid	*the extreme and mean ratio*
Romans	*aurea sectio (golden section)*
Luca Pacioli	*the divine proportion*
Christopher Clavius	*Godlike proportion*
Johannes Kepler	*the divine section*
Johann F. Lorentz	*the continued division*
J. Leslie	*the medial section*
Adolf Zeising	*the golden cut*
Mark Barr	Φ *(phi)*

Like God, the Divine
Proportion is always
similar to itself.

—Fra Luca Pacioli

Even as the finite
encloses an infinite series,
And in the unlimited
limits appear,
So the soul of immensity
dwells in minuta
And in the narrowest
limits, no limits inhere.
What joy to discern the
minute in infinity!
The vast to perceive in the
small, what Divinity!

—Jakob Bernoulli (1654–1705,
Swiss mathematician)

Mathematical formulas for Φ
are composed of an endless
chain of parts that resemble
each other and the whole.

under then over the ideal, getting closer and closer on its way toward an infinite at which it will never arrive. We watch as its decimals get longer and longer, like a thirsty taproot reaching for the infinite that beckons it onward. The farther out you go in the Fibonacci sequence the more precise the result becomes as it hones into the ideal.

The ideal it approaches has had many names over the centuries, often expressing highest regard for it with terms including "golden," "divine" and even "Godlike." It is the proverbial "golden mean," the ideal balance of life. Presently it is symbolized by the Greek letter Phi (Φ), named in this century to honor the Greek sculptor Phidias, who used it to proportion his designs, ranging from the Parthenon to his famous statue of Zeus.

Actually Φ is not a number but a *relationship*. We tend to focus on the visible numbers, but they represent the *accumulative process* that manifests them. We could actually start with *any* two numbers, not just zero and one, adding each consecutive pair to get the next term, yet the ideal limit will always approach Φ. Try it. Start with any two numbers and build your own sequence using the accumulative rule. Use a calculator to divide the consecutive pairs as fractions, and you will see the series also approach the ideal Φ. Different sequences become more exact at different rates.

Any way we approach Φ it leads us to its source in the infinite. Here are two equations mathematicians have found that also approach the ideal value of Φ. Don't be scared or anxious about them. Don't even try to solve these equations.

$$\Phi = 1 + \cfrac{1}{1 + \cfrac{1}{1 + \cfrac{1}{1 + \cfrac{1}{1 + \cfrac{1}{1 + \cdots}}}}}$$

$$\Phi = \sqrt{1 + \sqrt{1 + \sqrt{1 + \sqrt{1 + \sqrt{1 + \cdots}}}}}$$

Just *look* at them. Each is a picture, a mathematical mandala clothing the infinite as it zooms by. Notice how each is composed solely of unity interacting with itself, embedded Monads unfurling as far as we can see in a self-replicating rhythm, like mirrors facing each other whose reflecting image gets smaller and smaller. You don't have to see the whole to know it all. If you cut off a "branch" anywhere, the part resembles the whole equation!

We can make this relationship visible by geometrically dividing a line at its golden mean (follow the steps described with the illustration). When we cut any line we think we have produced two, a Dyad. But Dyads represent illusions.

Two things cannot be rightly put together without a third; there must be some bond of union between them. And the fairest bond is that which makes the most complete fusion of itself and the things which it combines; and proportion is best adapted to effect such a union.

—Plato

Whole

Φ (Phi)	Golden Mean
Large part	Small part

The golden mean is the only way to divide a line so that the whole and parts simultaneously relate to each other in the same way.

$$\frac{Whole}{Large\ part} = \frac{Large\ part}{Small\ part} = \Phi$$

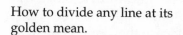

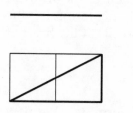 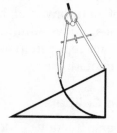

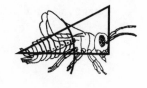

How to divide any line at its golden mean.

Start with any line and construct a square upon it. Quarter the square and draw a line from the corner to the midpoint of one side as shown, making a triangle. Put the point of your compass on the right end of the diagonal, and put the pencil on the square's bottom corner. Swing the compass up until it passes the diagonal (complete the circle). Then place the point of the compass on the other end of the diagonal and open the compass to the point just made. Swing the compass downward to see it cross the original line at its golden mean balance.

Φ in the pentagram. The pentagram star is itself a neatly packaged visual expression of the golden mean, everywhere crossing itself into Φ relationships. Each part relates to all others and to the whole in a chain of Φ proportions. When we look at any five-petaled flower we are seeing the flag of life, the self-reproducing force of life made visible as geometric harmony. In every flower we are literally gazing into the face of the infinite.

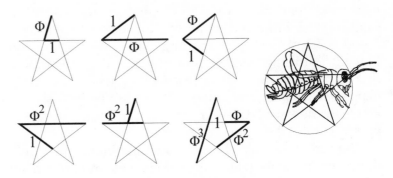

When one sees Eternity in things that pass away, then one has pure knowledge.

—Bhagavad Gita

Every cut actually creates a *three*-in-one Borromean bond between a large part, a small part, and the whole that unites them. A division at the golden mean brings these three parts into special harmony, proportioning them uniquely so that *the length of the whole line is to the large part in the same ratio that the large part is to the small part.* This is the beauty of the golden mean. It maintains a single relationship among the parts with each other and to the whole. This is the mathematical definition of harmony and balance, the truth of which nature embodies and we admire as her forms. It is the core idea of nature's self-replication and sits at the heart of the archetype of the Pentad.

Build Golden Mean Calipers

Enough with numbers, fractions, and formulas. We now know enough to construct a mechanical device that will locate the golden mean for us automatically.

Actually there are two devices; they are known as "calipers." They were used by Renaissance artists for laying out the proportions of their compositions on canvas and in stone. We can use them for exploring Φ relationships in geometry, nature, and art without resorting to numbers or calculation. Because a golden mean caliper is constructed on the frame of a pentagram, Φ is built into it. Mathematicians call this type of device a "linkage," or a "pantograph." It is used for making enlargements or reductions of a drawing by Φ proportions.

You'll need strips of firm material, like stiff cardboard or plastic straws, and metal brads, pins, or small bolts and nuts. If you have the skills and tools, finer models may be built of wooden or metal rods with bolts and butterfly nuts for easy loosening and tightening.

First, construct a very large pentagram star. Instead of an ordinary compass you may wish to use string attached to a tack and tied to a pencil.

Lay the strips (or rods) along the diagonals of the star as shown. Mark and cut the end points. Poke a hole at crossing points through which to insert the brads (or bolts). Notice that brads are placed as hinges at the vertex points and intersections of the star, so the leg must be slightly longer than the squared edges. Link the strips as shown in the illustrations.

The simpler golden mean caliper, made of two lengths joined at their golden mean, will indicate at its large side a distance Φ (1.618. . .) times larger than the distance between the pointers of its smaller side.

The second type of caliper will divide any *straight* line at its golden mean. Just put the extremes at the beginning and end of any line and its middle pointer will indicate the golden mean.

Both types of caliper will let you find the Φ proportions in the pentagram and decomposed pentagon.

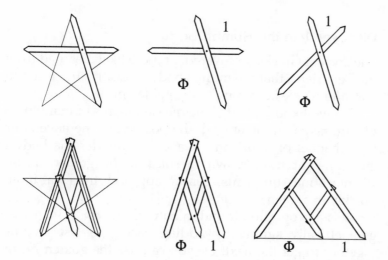

Golden mean calipers always show a changing, relative Φ relationship between pointers.

Golden mean calipers will reveal the Φ relationships in pentagonal geometry and nature's designs.

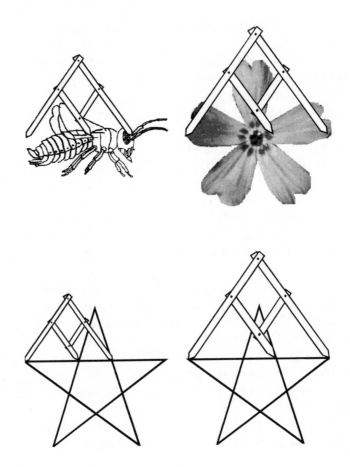

*H*armony makes small things grow. Lack of it makes great things decay.

—Sallust (c. 86–34 B.C., Roman historian and politician)

Explore Φ relations in the pentagram star.

*M*an is the measure of all things, of things that are that they are, and of things that are not that they are not.

—Protagoras of Abdera (c. 485–410 B.C., Greek philosopher)

Discover Φ in the Human Body

The ancient Greeks discovered, or more likely learned from the Egyptians, that the human body is ideally structured in part and whole according to the golden mean.

It's well known that if the measurements of many people are taken and averaged, the position of the navel, our first channel of nourishment and life, divides the body's entire height from crown to soles at the golden mean. Women's measurements seem to approach the ideal more quickly than men's.

Φ is a proportion found in natural growth. It increases geometrically simply by adding each two last lengths together to get the next. Or you can use the golden mean

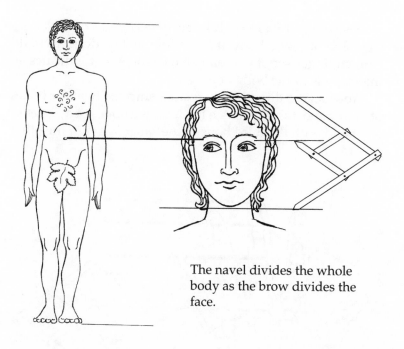

The navel divides the whole body as the brow divides the face.

calipers to extend a line into larger and larger Φ sections. This pattern is the frame of ideal expansion found in the whole and in great detail of the human body. Everyone is a little stretched or compressed from the ideal, but the statistical average tends toward Φ.

The basic unit is the vertical distance between the brow (the top of the eye) and the tip of the nose. The distance from the brow to the crown is Φ times larger than the brow-nose unit. The brow seems to be a turning point, a plane of reflection. The Φ ratio is also found as the distance from the nose to the base of the neck. It next extends from the neck to the armpit, there to the navel, to the reach of the fingertips, and finally from the fingertips to the soles. Thus, the body is ideally divided into seven Φ sections. This design was consciously used in ancient sculpture when depicting the ideal bodies of the eternal gods. Human figures were purposely depicted less perfectly.

The Φ proportion was applied in the sculpture of Phidias in various width relationships: head/throat, forearm/wrist, widest part of thigh/narrowest part of thigh, calf/ankle.

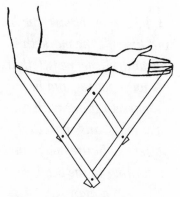

The wrist divides our cubit at the golden mean.

Use the golden mean calipers to verify these Φ relationships in your own body. It's possible, as I've done, to build a much larger set of golden mean calipers that indicates all these levels of the body at once.

You can see this additive relationship easily in the bones of your hand. Simply match the first two bones of a finger of one hand with the third bone of the same finger of the other

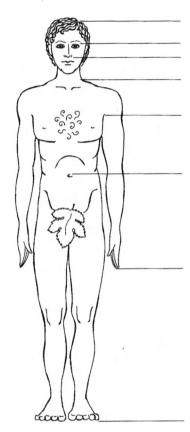

Ideal human proportions express a golden ratio sequence.

T he carpenter stretcheth out his rule; he marketh it out with a line; he fitteth it with planes, and he marketh it out with a compass, and maketh it after the figure of a man according to the beauty of a man.

—Bible (Isaiah 44:13)

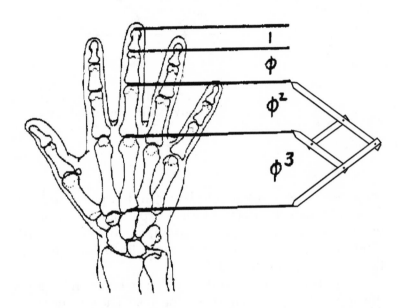

The hand begins to replicate the Φ progression of the whole body. In Arabic calligraphy, the shape of the hand spells "Allah."

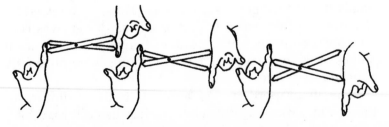

When the length of each finger joint is multiplied by 1.618. . . , the length of the next larger section is indicated.

hand. They'll be equal. And the second and third bones match the fourth bone within the back of your hand. In this way the proportions of the hand, like those of the face, replicate the scheme of the entire body. The parts model each other and the whole. The body is a repackaging of the pentagon, another whole in nature's fabric, with a flavor of the infinite.

We see the results of this self-growing symmetry in the beautifully expressive movements of dancers, the geometric precision of gymnasts, and the versatile, fluid motions of the hands of mimes.

Φ and the Body of the Solar System

A core theme in ancient philosophy is that of the human as a microcosm of the whole, the macrocosm. While the greater significance of that theme refers to correspondence with our subtle inner psychological and spiritual structures, it is also true to some extent for our physical configuration. We contain smaller similar structures and we are part of, and models of, larger structures of the cosmos.

Thus the human body itself models greater self-replicating wholes. Its expanding Φ proportion recurs in the larger structure of the solar system, in the distances of the planets from the sun and each other. Here the additive Fibonacci process is at work: the distance from the sun to Mercury plus Mercury's distance to Venus equals the distance between Venus and Earth. The solar system is modeled in the proportions of your body. The Φ ratio is most clearly defined by

As above, so below.

—Hermes Trismegistus (attributed to the Egyptian god Thoth, author of mystery teachings)

Man is all symmetry,
Full of proportions, one
limb to another,
And all to all the world
besides.
Each part may call the
farthest brother,
For head with foot hath
private amity,
And both with moons and
tides.

—George Herbert (1593–1633, English priest and poet)

The world is not a machine. Everything in it is force, life, thought.

—G. W. von Leibnitz (1646–1716, German philosopher and mathematician)

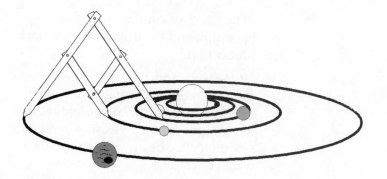

the planets nearest the sun, but the ideal gradually breaks down toward the outer planets.

In this way the solar system is a great body, perhaps a form of life we don't recognize as such.

The recognition of the Φ relationship structuring nature's forms gives us a passkey to our inner selves. The body's structure is a mirror of our psyche, a denser expression of the energetic patterns of our soul. Body and soul somehow partake of the same design. But in what way can a mathematical ratio permeate our souls? Through beauty. A deep part of ourselves recognizes in flowers and dancers the beauty of the mathematical infinite and sees in it the endlessness of our own depths. Natural beauty resonates with the archetypal nature within us. Let's see how the golden mean affects our sensibilities.

Beauty and the Best

Beauty is certainly a soft, smooth, slippery thing, and therefore of a nature which easily slips in and permeates our souls.

—*Plato*

What is beauty? It's something most people claim to be able to recognize but none can define to everyone's satisfaction. Philosophers and artists have tried:

Plato:	"The beautiful consists in utility and the power to produce some good."
	"The good, of course, is always beautiful, and the beautiful never lacks proportion."
Socrates:	"If measure and symmetry are absent from any composition in any degree, ruin awaits both the ingredients and the composition . . . Measure and symmetry are beauty and virtue the world over."
Aristotle:	"order, symmetry, and precision."

Quintilian:	"usefulness"
Thomas Aquinas:	"wholeness and harmony"
Leonardo da Vinci:	"basis in nature"
Michelangelo:	"Beauty is the purgation of the superfluous."
Emerson:	"The beautiful rests on the foundation of the necessary. The line of beauty is the line of perfect economy."
	"We ascribe to beauty that which is simple, which has no superfluous parts, which exactly answers its end."
Von Humboldt:	"Natural objects themselves, even when they make no claim to beauty, excite the feelings, and occupy the imagination. Nature pleases, attracts, delights, merely because it is nature. We recognize in it an infinite power."
Bragdon:	"The highest beauty comes always, not from beautiful numbers, nor from likeness to nature's eternal pattern of the world, but from utility, fitness, economy, and the perfect adaptation of means to ends."
	"There is a Beautiful Necessity which rules the world, which is a law of nature and equally a law of art."
J. G. Bennett:	"Not all forms of life can justify themselves on utilitarian grounds. We must here look at nature in her other aspects. She is beautiful and playful. To disregard these features is crass ingratitude."
Keats:	"Beauty is Truth; Truth, Beauty. That is all ye know on Earth and all ye need to know."
Confucius:	"Everything has its beauty but not everyone sees it."

*N*o *investigation can strictly be called scientific unless it admits of mathematical demonstration.*

—Leonardo da Vinci

Before you read any further, look at each of these rectangles and decide which is the most pleasing and which the least pleasing.

For quite a long time there have been attempts to establish a measurable link between inner and outer nature. In 1865 the controversial German physicist, psychologist, and philosopher Gustav T. Fechner (1801–1887) devised an experiment to give the "art" of aesthetics a mathematical foundation. He conducted one of the first public opinion polls to measure statistically aesthetic preferences (in Greek *aesthesis* signifies "sensuous apprehension"). He displayed ten rectangles and asked which was most pleasing and which least pleasing. His results, long since verified, were that people generally tend to prefer ratios and proportions closest to large consecutive Fibonacci numbers and thus the golden ratio. The "golden rectangle" is a rectangle whose sides are in Φ ratio. The one most like it, having sides in a twenty-one to thirteen ratio, is the fourth rectangle from the right.

Look around and you'll find many objects displaying Φ dimensions. Furniture, clothing, utensils, pens, mirrors, windows, writing pads, pans with handles, logo proportions, index cards (3 x 5 and 5 x 8 inches), grocery bags, picture frames, TV sets, cassette tapes, and stereo speakers often display the Φ ratio. One needn't be aware of mathe-

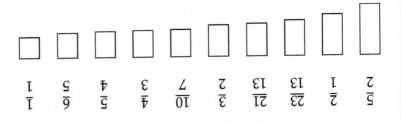

$$\frac{1}{1} \qquad \frac{5}{6} \qquad \frac{4}{5} \qquad \frac{3}{4} \qquad \frac{7}{10} \qquad \frac{2}{3} \qquad \frac{13}{21} \qquad \frac{13}{23} \qquad \frac{1}{2} \qquad \frac{2}{5}$$

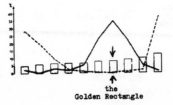

the
Golden Rectangle

Results of Fechner's survey. The golden rectangle was most often chosen as most pleasing and rarely chosen least pleasing.

matics to find this relationship pleasing. The golden rectangle appeals to us visually, but the proportion can also delight us audibly as in the building of syllables into words, sentences, passages, and books, from Virgil's *Aeneid* and in the rhythm of the chant "Kyrie eleison."

Φ's self-replicating symmetry appeals to us because we unconsciously sense its internal balance, recognizing in the harmony of Φ relationships the harmony within ourselves. Φ resonates with the core of life, reminding us of our own infinite depth and beauty.

Construct a Golden Rectangle

A golden rectangle is a rectangle whose sides are in Φ relation. To construct one, begin with a square and extend its bottom and top sides to the right. Place the unmoving leg of the compass at the midpoint of the bottom side of the square, and open the pencil to the upper right corner. Swing the compass downward until it intersects the extended bottom of the square (complete the circle for the full diagram). Do the same from the top midpoint of the square to the bot-

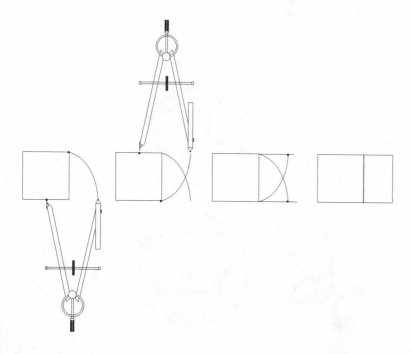

tom right corner. Connect the new points with a vertical line. In accord with the principle of self-similarity, the small, newly added rectangle is itself a golden rectangle, as is the whole rectangle including the square. That is, the area of the whole golden rectangle relates to the area of the square as the area of the square relates to the area of the smaller, turned golden rectangle. A three-in-one harmony of areas has been created.

Construct a 3-D Golden Frame

The self-similar harmony of the two-dimensional golden rectangle can become three-dimensional through the construction of a frame of three joined at right angles to each other. Upon this frame you can create the five Platonic volumes, the archetypal patterns of many minerals, plants, and animals.

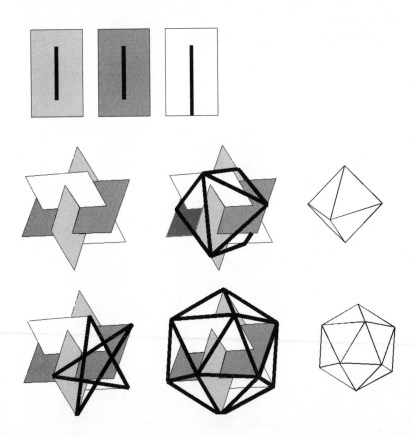

You'll need stiff cardboard (index-card material will work), a cutting tool, a needle and thread. Construct three identical golden rectangles. Cut slits along their centers as shown, long enough to slip another rectangle's smaller side through. The third rectangle, whose slit extends to one edge, slips around both to create the three-dimensional frame. You may wish to run tape along the joints to strengthen the frame.

Using a needle and thread you can sew through five corners to reveal the pentagram star (see illustration) that shares the Φ ratio with the golden rectangles. By connecting with thread only the midpoints of short sides, you can outline the edges of an octahedron. And by connecting every corner, you can make the twenty triangular faces of an icosahedron. With ingenuity, you can also build the remaining Platonic volumes on this frame.

Discover Φ in Classic Art, Crafts, and Architecture

From their deep study of geometry and nature the ancient mathematical philosophers influenced artists, craftspeople, and architects. Since ancient times the Φ proportion has

Regulating lines . . . are . . . a springboard and not a straightjacket . . . they satisfy the artist's sense . . . and confer on the work a quality of rhythm.

—Le Corbusier (1887–1965, Swiss architect, city planner, and writer)

Knowledge of a basic law gives a feeling of sureness which enables the artist to put into realization dreams which otherwise would have been dissipated in uncertainty.

—Jay Hambidge

This Babylonian stela depicting an initiate being led by priests into the presence of the sun god contains many Φ relationships.

This symmetry cannot be used unconsciously although many of its shapes are approximated by designers of great native ability whose sense of form is highly developed.

—Jay Hambidge (1867–1924, American architect, geometer, educator, and author)

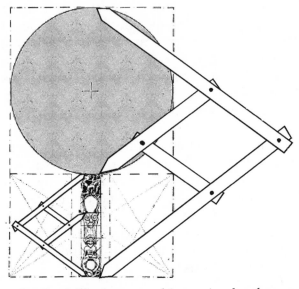

This circular Greek libation pan achieves visual and geometric balance with its handle by the golden mean. Together they fit within the frame of a golden rectangle divided into a square and smaller golden rectangle.

Art without knowledge is nothing!

—Jean Mignot (Gothic master builder from Paris, addressing architects in Milan, 1398)

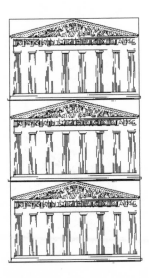

The west facade of the Greek Parthenon fits within a golden rectangle. The United Nations Building in New York was designed as three golden rectangles, like three Parthenons stacked upon each other, appropriate to the mission of promoting world harmony.

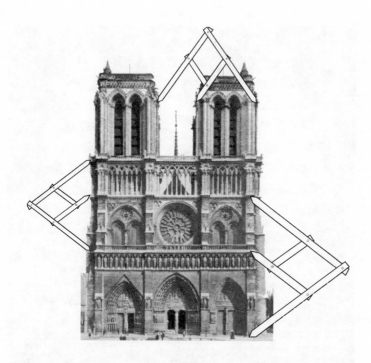

The west facade of the Cathedral of Notre Dame in Paris is rich in golden mean relationships. Use the golden mean calipers to verify them and find more.

shown up in artistic composition, it's simplest expression being a golden rectangle. The most complex designs employ the decomposition of the pentagon. Some designs are highly complex, such as those underlying Gothic cathedrals and paintings of the Renaissance. They are comprehensive and satisfying, even uplifting, due to the harmony inherent in the geometry of self-replication to which our eye responds.

Use the golden mean calipers to verify and explore the Φ proportions in the following compositions.

Discover Pentagonal Composition in Art

Of all peoples, the ancient Egyptians were the most brilliant at integrating geometry, symbolism, art, myth, and language in their works. For example, their hieroglyph for the *duat*, the "underworld" within ourselves, is a star within a circle. So whenever they created art that related to the *duat* they composed it on a large pentagonal grid decomposed into smaller pentagons and pentagrams. There are many examples of the Φ proportion in Egyptian art and architecture.

Sacred architecture is not, as our time chooses to see it, a "free" art, developed from "feelings" and "sentiment," but it is an art strictly tied by and developed from the laws of geometry.

—Fredrik Macody Lund

[The golden proportion] is a scale of proportions which makes the bad difficult [to produce] and the good easy.

—Albert Einstein

The gold mask of the sar-
cophagus of Tut-Ankh-
Amon, which contained his
mummy, was appropriately
designed using the self-
similar geometry of the pentagon, symbol of regeneration and rebirth.

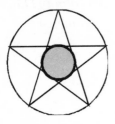

The hieroglyph of the *duat*
(Egyptian for "underworld")
underlies the geometry of
Egyptian art concerning jour-
neys through the under-
world.

Pentagonal symmetry allows each small part to be a
whole in itself while harmonizing with other parts and the
greater whole. The visual effect is pleasing, balanced, even
beautiful, and conveys the star's feelings of excellence,
power, authority, life, and humanness.

Use the golden mean calipers to find the Φ relationships
harmonizing the elements within each composition.

With the publication in 1509 of Fra Luca Pacioli's *De Divina Proportione,* which marveled at the geometric relationships within the pentagon, artists regained access to the
star's construction and to its value for organizing space aes-
thetically in harmonic composition. As Rome's director of
antiquities, Raphael (1483–1520) was responsible for drawing newly unearthed art, crafts, and architecture. He noticed

Salvador Dali was a student of ancient design and used pentagonal symmetry in many of his paintings.

Jewelry of Tut-Ankh-Amon from Thebes. The rectangular pectoral symbolizes the creation of the universe by the sun above the waters of chaos, while the nearly triangular pectoral represents the birth of the sun and moon. Both were designed using pentagonal symmetry. Note how the sun in one and the moon in the other are at the center of five-pointed stars, forming the glyph of the *duat*, the "underworld" path to rebirth and regeneration.

*T**he good, of course, is always beautiful, and the beautiful never lacks proportion.*

—Plato

*B**eauty doth of itself persuade the eyes of men without an orator.*

—William Shakespeare

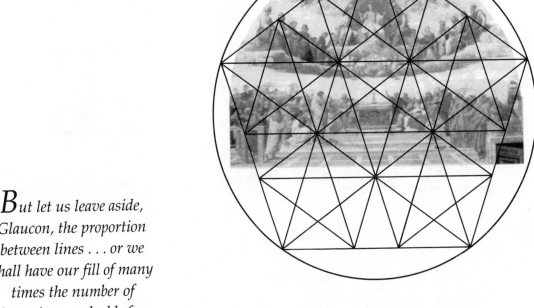

*B*ut let us leave aside,
Glaucon, the proportion
between lines . . . or we
shall have our fill of many
times the number of
discussions we had before.

—Socrates (470–399 B.C.,
Greek philosopher)

that the ancient Roman works emulated the Greek and Egyptian symmetries. From this observation and his many teachers, Raphael learned to use the pentagon's compositional harmony which he, like many artists of the Renaissance, applied to quite a few of his paintings. His *Dispute over the Holy Sacrament* uses the regenerating pentagram inscribed in the circle. The subdividing pentagons and pentagrams provide a frame of regulating lines along which both the figures and architecture are aligned.

○

Our final look at the principles of the Pentad draws us into the most widespread shape in nature—the spiral.

Spirals are deeply rooted in the architecture of the universe; they are found in every size and substance. We're always intimate with spirals yet rarely notice them. Sometimes we miss them due to familiarity, as in water whirling down the tub's drain and in the shape of our ears. Sometimes we miss them because of their obscurity, as in the spiral "staircase" of leaves whirling around a stem. Sometimes we miss them because of their size, or distance, hurricanes or galaxies. And sometimes we miss them because of their invisibility, as in the shape of the wind and waves of emotion.

What makes spirals so prevalent in cosmic design? They are the purest expression of moving energy. Wherever energy is left to move on its own it resolves into spirals. The universe moves and transforms in spirals, never straight lines. Spirals show up as the paths of moving atoms and atmospheres, in molecules and minerals, in the forms of flowing water, and in the bodies of plants, animals, humans, and the greater bodies of outer space. A universal integrity of spirals unites all creation.

Like the pentagram star, the spiral expresses the geometry of self-similarity. At first, the spiral doesn't appear to be pentagonal, but wherever you see a star you will find a spiral rolled within it. When we feel its hypnotic lure while gazing into a spiral we are responding to the Pentad's siren call into the self-similar infinite.

The spiral's role in nature is transformation. Similarly, in myth and religion it is the path of spiritual and mystical transformation.

You may be surprised at all the spiral processes you already know about; now you may learn to look at plants, fruit, vegetables, and other familiar natural forms differently.

Three principles of nature's spiral will reveal to us its secrets of universal construction.

- Spirals grow by self-accumulation.
- Every spiral has a "calm eye."
- Clashing opposites resolve into spiral balance.

UNRAVELING SPIRALS

Growth by Self-accumulation

Although mathematicians distinguish many species of spirals, they all have in common the fact that they wrap around a fixed point at a changing distance.

We can begin to understand spirals by doodling them. Draw some spirals with your normal writing hand. Then draw spirals with the opposite hand. Then read on.

Did your spirals spin in the same or opposite directions? Did they flow outward or inward? Did the distance between coils stay the same, widen, or narrow?

Adding to the definition of these shapes, we can say that different spirals grow by curling around their points at different rates. The two main types of spirals we're most often exposed to are called the Archimedian spiral and the golden spiral.

Named after the Greek mathematician and scientist Archimedes, who described it in his book *On Spirals* (c. 225 B.C.), an Archimedian spiral is one whose distance from the point grows at a fixed rate. The distance between successive coils is always the same. We encounter the Archimedian spiral as a coil of rope, clock springs, record grooves, and a roll of paper towels. The helix is a three-dimensional version of the Archimedian spiral, found in bolts and coil-springs, in candy-cane stripes and barber poles, and in the double helix of the DNA molecule, our genetic heritage.

But the spiral most commonly found in nature's public manuscript is another type, the golden spiral. Unlike the Archimedian spiral, the distance between the golden spiral's coils keeps increasing, growing wider as it moves away from the source or narrower as it moves toward it. Called by many names through history, it was apparently well known in ancient times as evidenced in prehistoric art and worldwide ornament. This is nature's spiral of seashells and ram's horns, of our ears and fists, of watery whirlpools and star-spangled galaxies. It grows from within itself and increases according to the Fibonacci process of accumulation. A simple experiment will reveal the important difference in the ways these two kinds of spirals grow.

It is perhaps a more fortunate destiny to have a taste for collecting shells than to be born a millionaire.

—Robert Louis Stevenson (1850-1894, Scottish essayist, novelist, and poet)

Fold the Archimedian and Golden Spirals

Cut two strips of paper. On one, mark off distances at one inch (or one centimeter, or one finger width, or any unit you may choose), at two inches, three inches, four inches, and so

1	2	3	4	5	6

1	1	2	3	5	8

Archimedian spiral

Galaxy

Nautilus shell

Golden spiral

Red cabbage

Whirlpool

Rope

Clock spring

Paper towels

on. In other words, make the counting sequence visible. Fold a right angle at each mark so that the strip turns around to become an Archimedian spiral. The distance between its coils remains constant, always that of a single unit. Its accumulation comes from outside itself, always adding an external "one" to the previous term.

On the second strip, mark off units measured by the Fibonacci sequence 1, 1, 2, 3, 5, 8, 13, 21, . . . inches. When you crease and fold the strip you'll see that the distance between its coils continually increases. This is the golden spiral found in nature. Like the famous Fibonacci number series, the golden spiral *grows from within itself*; nothing is absorbed from outside. It is the physical representation of self-accumulation. Its appearance tells us that the Fibonacci sequence and Φ underlie an internal harmony, excellence, and dynamic balance at work.

○

The Swiss mathematician Jakob Bernoulli (1654–1705), patriarch of a family of distinguished mathematicians and scientists, devoted a great deal of study to this particular spiral. He discovered its self-accumulating, self-reproducing nature and gave the spiral a motto (perhaps the only one associated with a geometric shape): *Eadem mutato resurgo—* Although changed, I arise again the same. He was so impressed with the properties of the golden spiral that he requested that the shape and its motto be carved onto his tomb. Unfortunately, the stonemason mistakenly carved an Archimedian spiral. Perhaps he didn't realize the important difference between them, or perhaps he just didn't know how to geometrically construct the spiral Bernoulli had requested. Here's how it's done:

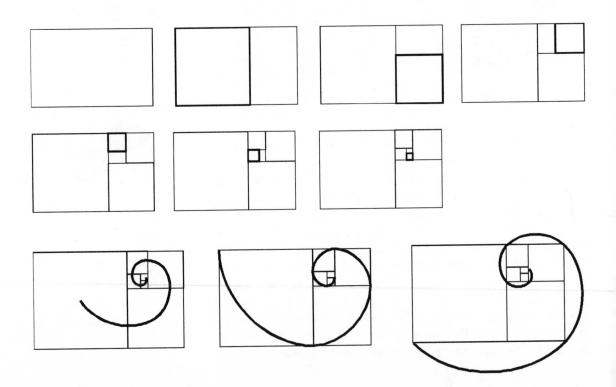

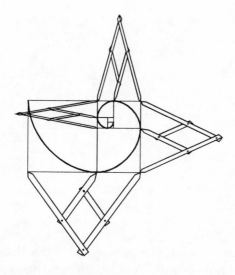

Unfurl Nature's Golden Spiral from a Golden Rectangle

The golden rectangle, whose long and short sides are in a Φ relationship (or approximated by large consecutive Fibonacci numbers), holds nature's spiral coiled within it. After constructing a golden rectangle of any size, put the point of your compass on a corner and open it to the length of the short side. Swing the compass to measure this distance onto the longer side. Turn the compass around, and swing it to mark off the same distance onto the other long side. Connect the points to make a square within the golden rectangle. Astonishingly, what remains is a smaller golden rectangle turned on its side.

If this process of marking out a square from each new golden rectangle's short side is repeated, a smaller, turned golden rectangle will always remain. Lop out a square within it and another golden rectangle appears. This process of squares chasing a single receding golden rectangle is mathematically endless. Theoretically, there is no end to the trail of squares cut out of a continually resurrecting golden rectangle, seducing the squares forever onward. Connect corresponding points on each successive whirling square along their centers, internal corners, or external corners and the spiral appears as a smooth curve. Life's proportion lures

Subtracting whirling squares from within the small side of a golden rectangle shows how natural forms like whirlpools "dissolve" by removing self-similar parts. Adding squares onto the large outside of a golden rectangle shows how growth occurs through the accumulation of self-similar parts.

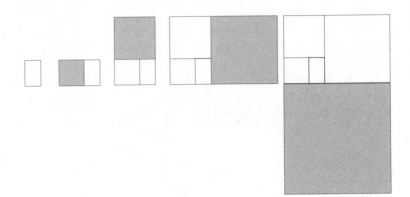

the square ahead along a spiral. This diminishing path shows us how events grow smaller and *dissolve* the universe, like water whirling down a drain or electrons spinning out in a bubble chamber.

Growing larger, the spiral exhibits *expansive* growth in the form of seashells and hurricanes. We can expand a golden rectangle by adding a square onto its larger side. The result is another but larger self-similar golden rectangle. By continually *adding* another square onto the outside of the larger side of each successively produced golden rectangle, another golden rectangle will hurl toward the larger infinite in perfect Φ proportion.

The spiral results from the play of the rational square with the transcendental Φ ratio. It is the play of life itself, clothed by the four states of *mater* that make it visible. As it configures, the mysterious force of life leaves a spiral trail of matter, symbolized by the squares left behind in its wake.

Discover the Spiral in the Star in Plants

The spiraling processes of self-accumulation and dissolution are not limited to rectangles and squares. Triangles whose larger and shorter sides are in a Φ relationship, golden triangles, also self-replicate, whirling within and around one another as the same golden spiral of nature.

Every triangle in a pentagram is a golden triangle. Thus, when we see a five-petaled flower we can always find a spiral associated with it, as the buds slowly whirl open.

You can see the spiral and star simultaneously by look-

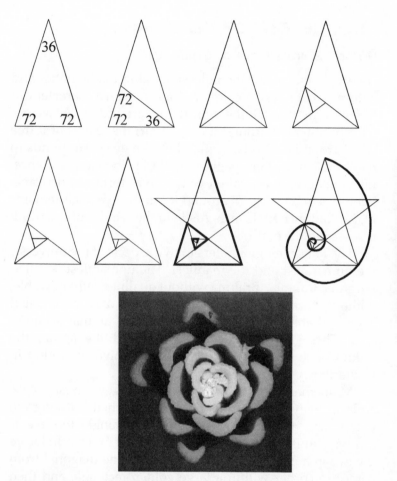

The self-similar logarithmic spiral is found in the self-similar parts of a golden triangle.

A "sun's-eye view" along the stem of a weed or any plant or tree reveals its whirling pentagonal structure.

This slice of a celery bunch shows a spiral arrangement of stalks accumulating as a pentagram star.

ing at plants from the "sun's-eye view" straight into its stem or branch. Common weeds are excellent for this. Squint a little and you'll see the irregular overall fiveness of the leaves and branches as they wind up the stem. If they formed a perfect pentagram the leaves would overlap. But a slightly irregular star shape permits each leaf to be exposed to the sun.

Peel back the leaves of a cabbage, lettuce, or any plant in the reverse of the sequence in which they grew and you'll see that they form the points of a stilted pentagram star.

Spiraling bracts of a pinecone.

INTO THE "EYE" OF THE STORM

Natural Spirals Have a "Calm Eye"

A curious fact about the golden spiral in mathematics and nature is that it has a "calm eye," a core with a character different from that of the whirlwind around it. A watery whirlpool wraps around a core of air; the leaves of a tree whirl around a woody trunk. The calm eye corresponds to the "zero" at the very beginning of the Fibonacci sequence. Like the deceptively calm eye within a storm or hurricane, intense action swirls all around it but the eye itself remains placid, untouched. Leaves rustle in the wind, but the trunk is rooted in the earth.

The Hindu religion refers to two aspects of the universe, the unmanifest and the manifest. The unmanifest is called Nirguna Brahma, "Brahma without qualities," unknowable, without characteristics, while the manifest universe is called Saguna Brahma, "Brahma with qualities," all that is knowable. These correspond to the eye and the spiral, the unknowable source and the turbulent universe around it, mysteriously driven by the "calm" core.

Mathematicians call the golden spiral's eye an *asymptote*, a place always approached but never reached. Although to the squares whirling in the golden rectangle the eye is always infinitely ahead, we can pinpoint it exactly. To locate the eye in a golden rectangle just draw one diagonal from corner to corner within a large golden rectangle and then another diagonal in the next smaller golden rectangle. The eye is located at the point where the diagonals of every golden rectangle cross.

Leonardo da Vinci wrote, "A vortex, unlike a wheel,

Whirlpool funnel

Daisy

Fingerprint "plain whorl" type spiral

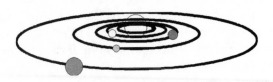

The planets roll on the curved surface of a whirlpool called the solar system with the sun at its eye.

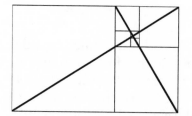 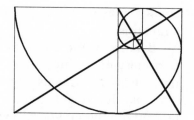

Finding the spiral's eye in a golden rectangle.

Every storm hath his calm.

—Robert Greene (1560–1592, English poet and playwright)

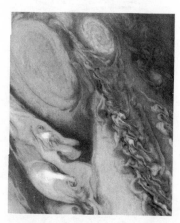

Jupiter's "eye" is a spiraling storm many centuries old.

moves faster toward its center." He was referring to a fascinating difference between these two forms. Watch a bicycle wheel turn: you can see the spokes move *more slowly* near the hub than at the rim. A spiral does the opposite, moving *faster* at its eye than it does further out. The solar system is a great whirlpool with planets rolling on its surface. Those closest to the sun orbit faster (Mercury in eighty-eight days) than those at the periphery (Pluto in 248 years). In miniature we see this characteristic in water whirling down the drain and in low-pressure storms twisting the atmospheres of many planets. If you watch a daisy flower grow over time you'll see its central yellow florets ("little flowers") emerge mysteriously from the living eye of the plant and grow, whirling outward more slowly toward the periphery. In fact, the name *daisy* derives from "day's eye," the eye of the day.

Learn to recognize spirals in nature, in the living whirlpools we call "trees" and in house plants. Look for the "eye." Whatever its substance, the eye is always mysterious and hypnotically fascinating to watch and to contemplate as the source of the spiral.

In watching water whirl down a drain, note that the narrower and calmer the eye is, the greater the speed and turbulence around it. Notice also that unlike the center of a wheel, the spiral's eye is not fixed in one place but is dynamic and flexible.

Far from insignificant, the calm eye is the spiral's center of gravity around which it all balances. Without the eye there would be no spiral expansion or dissolution, no whirling, no balance, no life. A spiral is balance-in-motion made visible, a graph of forces displaying change without

Underwater view of a whirlpool's flowing circuit.

The digestive system of a pig, like that of other animals, follows a whirlpool to transform food into energy and waste.

change, playing out before our eyes. Two activities demonstrate this:

Balance-in-motion: Drop thread or a toothpick near the periphery of a whirlpool of water going down a drain and notice how as it travels around *it continues to point in the same direction.* Its speed changes, but its orientation doesn't (although when it goes down the eye it will point upward). This expression of its mathematical self-similarity explains why each planet whirling around the sun keeps its axis pointing in the same direction throughout its orbit. The Earth's axis always points 23.5 degrees off the "plane" of the solar system, while Uranus rolls on its side at 89 degrees, its poles pointing toward the sun twice yearly. This is the kind of "change without change" that inspired Jakob Bernoulli's epitaph.

Self-replicating balance: Another way to see this principle of balance through change is to bend a piece of wire into a golden spiral and balance it by a pencil or straw through the eye. Observe how it settles into a state of balance with its end pointing upward. Then, with a wire-cutting tool snip a length off the end and observe the spiral shift to rebalance itself. Remarkably, the "new" spiral settles to resemble a miniature model of the original spiral's orientation, maintaining change without change.

We can also understand the spiral's balance by viewing it in sections. Although different in size, every segment has the same curvature. If a microphotograph into the eye were enlarged, it would find another part of the spiral upon which it would fit exactly. No other spiral will do this, nor

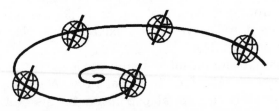

Change without change. An object swirling around a whirlpool remains pointed in the same direction.

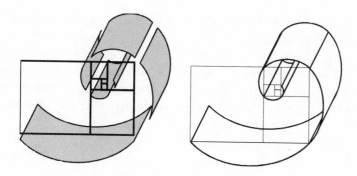

will any other spiral accommodate the dynamic balance that nature values. An Archimedian spiral's curvature gets broader as it widens, no section being identical with any other.

These characteristics are the open secret of balance of animal horns, seashells, plants, and galaxies. For example, as the chambered nautilus creature grows larger, the gland that exudes shell material also grows, building a widening shell. The shell's golden spiral shape maintains the same center of gravity in any size, and so the nautilus need not relearn how to balance itself as it matures. The same is true for the growing horns of a ram. As the horn material accumulates, growing larger and more massive, its golden spiral maintains the same center of gravity. Thus the ram need not adjust its posture throughout life to uphold the growing horns. Similarly, the tree that puts out branches and leaves in spiral "staircases" around their respective "eyes" can get enormously large; yet the tree always balances no matter how massive it grows to be.

In a whelk shell, the spiral's eye is the central column that gives the whole shell outstanding strength. Gothic and Renaissance artists used local shells to design spiral staircases of great strength. The well-known sound of the ocean "heard" in a spiral seashell is actually the magnified echo of the blood pulsing in the listener's own ear.

Divers observe that schools of fish, like groups of birds and swarms of insects, regroup in the pattern of a whirlpool spiral. Some say that this is because they are following the

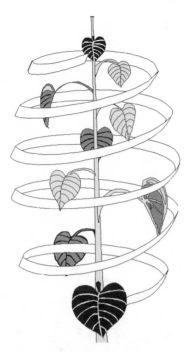

The leaves of a plant form a continuous spiral around its stem (eye), the balanced structure of a whirlpool.

unseen whirling water or air. Others maintain that each individual fish, bird, and insect is like one cell, one part of a single organism and they are linked as one energy field.

Wherever the spiral appears in plants, animals, or solar systems, it insures dynamic balance during inevitable growth and dissolution. Learn to identify and then contemplate the eye in natural spiral phenomena.

Another property unique to this type of spiral is reflected in an alternate name for it, the "equiangular" spiral, coined by the French mathematician and philosopher René Descartes in 1638: any line drawn from the pole to the curve cuts it at the exact same or "equal" angle all around.

The flight of a fly illustrates this. A fly's eye is made of thousands of tiny, hexagonally packed facets. It sees the same tiny image in each facet, instead of one large image as we do. As the fly approaches an object, the image shifts slightly in each facet. In response, the fly shifts its body so that the image remains fixed in its original place in each facet. As the distance to the object becomes shorter, the angle of approach remains constant. Thus, the path produced is

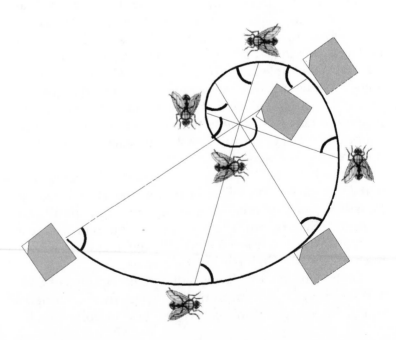

A fly approaches an object along a spiral path due to the structure of its eyes and nervous system.

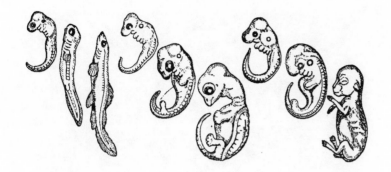

Fish, tortoise, rabbit, and other vertebrate embryos begin life as unfurling spirals.

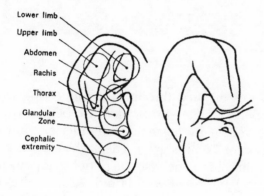

Lower limb
Upper limb
Abdomen
Rachis
Thorax
Glandular Zone
Cephalic extremity

Acupuncture makes use of the way the spiral human embryo resembles the spiral of the ear.

made of segments whose arcs are identical, only different in size—tracing the accumulative golden spiral.

Like other vertebrates (creatures with bones) we spin into life in the womb as a spiral embryo. We are conscious energy systems that whirl into configuration as other natural energies do along spiral lines of force, unfurling into life along an endless path of spiral transformation.

The outer ear, or auricle, is one of our body's most obvious spirals, along with curls of hair and spiral cowlicks, the mushroomlike vortex of our nostrils, our curling fists and accumulating fingernails. Chinese acupuncturists, recognizing the principle that the part models the whole, map the entire spiral embryo onto the spiral ear. Thus, to treat any malady of the body, they'll first treat the corresponding part of the ear.

A man is the whole encyclopedia of facts. The creation of a thousand forests is in one acorn, and Egypt, Greece, Rome, Gaul, Britain, America, lie folded already in the first man.

—Ralph Waldo Emerson

The outer ear and inner ear (cochlea) are spirals. The shape of the cochlea, the organ that "hears" music, corresponds to how chromatic musical octaves appear when graphed as wavelengths. Each note is identical to those directly above and below it on the spiral but with a one-octave difference.

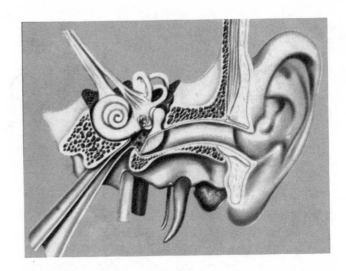

The man who has music in his soul will be in love with the loveliest.

—Plato

Only in this century, through the pioneering work of Georg von Bekesy (1899–1972), have we begun to understand how we hear by means of the inner ear, our spiral *cochlea* (Greek for "snail shell"). The cochlea of each ear is encased within a hard shell (so we're not overpowered by our own voice) and filled with liquid. Around the central column is a "handrail" of sensitive hairs that decreases in thickness as it winds upward. Sound vibrations enter the whirlpool of the outer ear and tap at the cochlea's base, traveling in waves up and around the inner liquid. Each tone makes waves that break at a different point up the handrail according to the tone's wavelength. The resulting spiral turbulence of the breaking wave triggers the sensitive hairs at that location. The hairs then transduce the mechanical waves into unique electrical signals sent to the brain for interpretation.

Humans hear approximately ten octaves of sound in a cochlea of two-and-three-quarter spiral turns. The cochleae of other mammals have various numbers of turns and extended opportunities for finer waves to break, which explains why dogs and mice can hear frequencies higher than humans can perceive.

The first recorded fact of mathematical physics, attributed to Pythagoras, is that musical notes can be expressed as

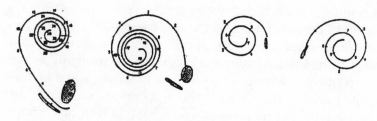

The cochlea of a cow, guinea pig, mouse, and rat are spirals that register different levels of frequency.

mathematical ratios (see chapters seven and eight). So when the notes of the chromatic musical octave are graphed according to their frequency of vibration, as accumulating wavelengths, the graph forms the shape of the golden spiral. Stretched up into three dimensions it becomes the shape of the cochlea, the organ that transduces sound to consciousness. The cochlea and all nature's spirals can thus be seen as solid, liquid, gaseous, and fiery representations of the music that structures the cosmos.

Why do waves break in the cochlea where they do? And exactly how *do* curling waves rise from the sea and break onto the shore? This bit of wisdom belongs to the lore of surfing knowledge and what is referred to as "shooting the curl," or skimming through the wave's eye. Winds blowing onto the oceans cause its huge, rolling waves. The sharp eye of the surfer notes the *distance between* successive rows of

Ye waves, that o'er the interminable ocean wreathe your crisped smiles.

—Aeschylus

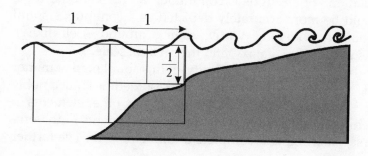

Behold the Sea . . . [in its] mathematic ebb and flow, giving hint of that which changes not.

—Ralph Waldo Emerson

Spiral heart muscle

incoming waves, their "wavelength." When the narrowing depth to the bottom below the wave is *exactly half the wavelength* between the waves, it will rise, push forward, and meet air resistance, causing it to curl around itself into its eye as an upright whirlpool to form a breaking wave of fine white bubbles. Why does it rise and curl at exactly *half* the wavelength? For this reason you'll have to wait until chapter eight and the discussion of the "doubling" principle of musical octaves.

Not only does our whole body whirl into life as a spiral, but at its heart, our body's *literal* heart, is a dexterous (left-spinning) spiral vortex. Sandwiched in the middle of its three layers is a sheet of contractor muscle in the shape of nature's familiar golden spiral. Like the sun at the center of our personal solar system, this living whirlpool, the size of your spiral fist, contracts and relaxes with great strength as the familiar "lub-dupp" beat sending blood rushing from its left ventricle chamber (a smaller spiral whirlpool within it) through 60,000 miles of blood vessels. The inner balance we seek, therefore, is already represented within the very fibers of our heart.

○

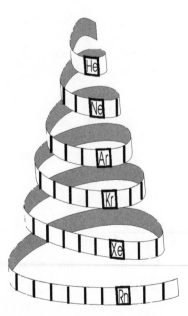

The Periodic Table of Elements as a continuous spiral.

For over a century scientists have known that the properties of the ninety-two natural atoms, and nearly a score of artificial ones, repeat in cycles or periods, like the notes of the musical octave. Chemists have arranged this cyclic list of atoms into the well-known Periodic Table of Elements, usually represented as a flat chart in rows and columns showing accumulations of electrons and protons, with their atomic numbers from one to ninety-two and beyond. But the "accumulation" of electrons is continuous, so the Periodic Table would be more accurately depicted as a continuous spiral ribbon, a widening vortex rolling around itself like a seashell, cochlea, galaxy, and the musical scale.

The eye, or center of gravity, of this spiral map of matter falls along the column of inert gases, sometimes called noble gases, because of their disdain for joining other elements in chemical bonding. They are the stable elements of the universe, balanced and complete unto themselves. The farther

Symbol	Name	Atomic Number	Closest Fibonacci Number
He	Helium (*sun*)	2	2
Ne	Neon (*new*)	10	8
Ar	Argon (*inert*)	18	21
Kr	Krypton (*hidden*)	36	34
Xe	Xenon (*stranger*)	54	55
Rn	Radon (*ray*)	86	89

an atom falls on the Periodic Table from an inert gas, the more willing it is to seek completion by entering into a chemical relationship with other atoms. The closer the elements are to an inert gas, the less needy they are.

A look at their atomic numbers shows the inert gases to have quantities of electrons very close to numbers from the Fibonacci sequence. If a spiraling strip of paper representing the Periodic Table whirlpool were marked off at Fibonacci lengths, the markings would fall on the approximate locations of the inert elements, its calm core.

All the matter of the universe, represented by the Periodic Table of Elements, resembles the spiral structure of smaller whirlpools that comprise this whole. Even the tiniest subatomic "particles" are vortices that reveal their spiral spin in bubble chambers. From the smallest particle whirlpools to the largest clusters of galaxies, each dances the same spiral step when coming into, or dissolving out of, manifestation.

A single electron in a super-heated bubble chamber leaves spiral trails as it loses energy within the chamber's magnetic field.

THE DANCE OF SPIRALS

When Opposites Clash They Resolve into Spirals

What brings spirals about? In what conditions do they arise? Spirals are a sign of growth and transformation through resistance. While their sizes and substances vary widely, nature's spirals result from an interplay of opposites, the clash of the Dyad. In its geometric construction, we saw the spiral arise at the interface, where rational squares endlessly chase their opposites in the transcendental golden rectangle.

*A*tom *from atom yawns as far As Earth from Moon, or star from star.*

—Ralph Waldo Emerson

The current that with gentle murmur glides, Thou know'st, being stopp'd, impatiently doth rage.

—William Shakespeare

A jet of milk squirted into water resembles mushroom growth.

The interface where hot meets cold, motion meets stillness, rising crosses the path of falling, any of the dual throng will clash but resolve into a dynamic spiral balance. Even love and hatred, ignorance and wisdom find balance in the spiral dance of life.

Four types of spiral appear in nature: the whirlpool eddy, the wave, the mushroom vortex rings, and the vortex street. Actually, each of these is a variation on the fundamental whirlpool "eddy" (in Old Norse *idha* signified "that which flows back around").

The Mushroom Vortex

This principle of opposites resisting and resolving into spirals is most easily understood as the mushroom vortex, or vortex rings. You see them when you pour a moving liquid like milk into a motionless liquid like coffee or tea. The poured liquid meets resistance at its head and curls to the side. Each turn also meets resistance and curls further inward, repeating the cycle of meeting resistance and curling until the force dissipates and resistance disappears, leaving only the spiral path. The result appears in the form of a mushroom, which should be no surprise, since a living mushroom is mostly water, too, although it moves more slowly while displaying the same pattern as clashing liquids do.

A reversal of mushroom vortex rings, called "backwater vortices," forms behind a flat plate as water or air rushes by it. These vortices are studied in wind-tunnel tests for cars, boats, racing bicycles, and airplanes, where the spirals are known as "turbulence." If you've ever driven behind

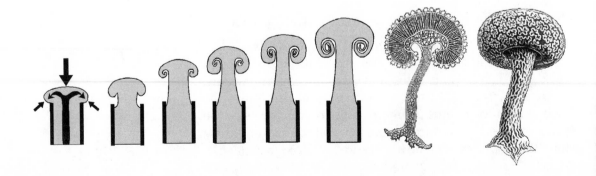

another vehicle on a dusty road you may have noticed the dual mirror spirals trailing the moving car as it thrusts through the still air. Dust makes the spirals visible.

We can see that the mirror spirals of the bivalve clamshell, as well as the creature within, and leaves display mushroom vortex rings. Only this dual golden spiral will allow each half to accumulate material on its lip and grow larger in mirror fashion without interfering with the other half's increase.

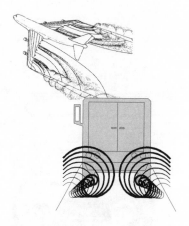

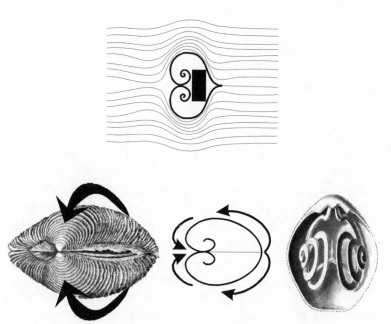

Bivalve shell grows as mirror spirals.

Many leaves unfurl as mushroom vortices, which the ancients saw as the nostrils of Pan, the spirit of the living forest.

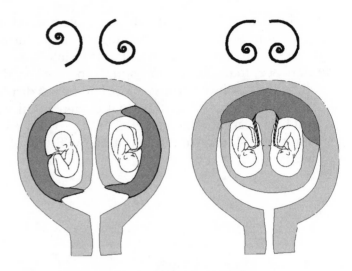

Fraternal and identical twins as mirror spirals.

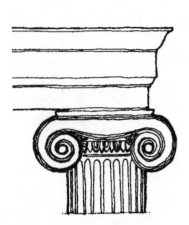

The capital of a Greek Ionic column symbolizes the Earth's subtle energy flowing through the temple, meeting resistance at its head, and rolling back as mushroom vortex rings.

The mushroom vortex also describes the entry of twins in the womb, who whirl into life together. Since each must not interfere with the other's growth, they develop as vortex rings, growing simultaneously larger while each maintains its unique center of gravity. Fraternal twins arise from two fertilized eggs, while identical twins are born of the same egg which has split. The spirals of fraternal and identical twins mirror each other differently.

The Vortex Street

Better to hearken to a brook, than watch a diamond shine.
—*George MacDonald*

If you've ever been in a rowboat you may have noticed whirlpools trailing behind the moving oar as it cuts through still water. Look closely and you'll see that the eddies are linked in a daisy chain of alternately opposite spinning whorls known collectively as a "vortex street." The phenomenon also occurs behind a still branch dipping into a moving stream.

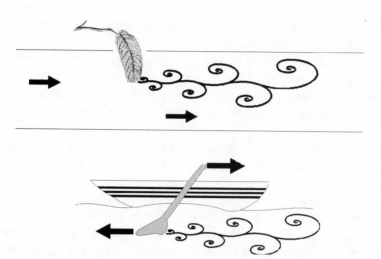

When a still object interrupts a moving stream, or a moving object disturbs still water, a vortex street forms to reestablish balance.

These phenomena in water and air may, we think, be regarded as the letters of a script which it is necessary to use like the alphabet of nature. Those who wish to remain at the stage of pure phenomenology relinquish the ability to read this writing and thus the ability to understand its meaning. They see the letters, but not words or sentences.

—Theodor Schwenk

The central axis of a vortex street is a forward-flowing zigzag from which the separate vortices emerge and balance. The central rhythm of alternating pulsation gives the whole "street" stability. Each spiral whirls independently balanced, and all spin in directions that support the direction of the central stream. To understand its rocking rhythm, think of how you break into a run, ride a bicycle, or roller-skate: you start off rocking side to side to achieve balance while thrusting forward. The more stable you become, the less polarized and straighter the central axis becomes. This is how the archetype struggles with resistance to resolve itself into a state of dynamic balance.

The moving wings of a jet plane split the still air and leave a vortex street of spiral turbulence behind them. Designers know that turbulence is a cause of drag on the vehicle, and they strive to reduce it.

Flocks of flying birds take advantage of vortex streets in their familiar V formations. Only the lead bird must really work at flapping its wings; the others latch onto the undulating spiral wake of turbulence trailing behind it. They simply relax their wings and let the rolling waves move them up and down and forward. When the lead bird becomes

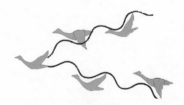

Birds rely on vortex streets to make long-distance flights easier.

tired she falls back while another moves ahead to work at splitting the breeze for the others.

The sound made by the wind is a measure of its turbulence. The wings of some birds like owls are so perfectly designed that they produce virtually no turbulence, and consequently no sound, enabling them to silently surprise their prey.

Anything that splits the breeze, from your finger to a tree or a building, will cause an invisible vortex street to spread behind it.

Due to the spin of the Earth and its Coriolis effect (deflection of a moving object), low-pressure storms north of the equator spin counterclockwise, meshed with high pressure spirals (fair weather) spinning clockwise. For the same reason, directions are reversed south of the equator. To see the Coriolis effect in miniature, that is, to see how the Earth's spin causes the atmosphere to spiral, shoot a marble from the rim of a spinning record turntable toward its center (as if above the South Pole). The marble's path will curl as a golden spiral like the great winds.

Observe a weather map and you'll see that high- and low-pressure spirals are linked as a daisy-chain vortex street in bands around the planet. These "jet streams," high-altitude zigzagging winds, are the central lanes of great vortex streets meandering between whirling atmospheric spirals. Global weather is all meshed as such spirals. The Earth's atmosphere is one whole. Changes in weather patterns anywhere affect weather everywhere.

. . . And in the shifting of the winds, and in the clouds that are pressed into service betwixt heaven and earth, are signs to people who can understand.

—Koran

Sand dunes are carved by one side of the moving vortex street we simply call "the wind." Similarly ribbed sand at the shore is carved by the vortex street formed by the ebbing water clashing with the incoming wave.

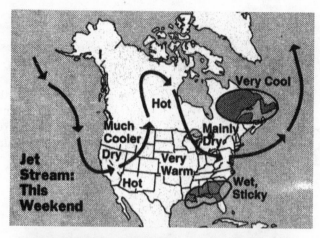

The jet stream is the central zigzag of a vortex street garlanding the planet. The spirals that spin at its sides are the meshed high- and low-pressure systems of our weather.

Spiral storm

... $t$he babbling gossip of the air.

—William Shakespeare

Differences in temperature, motion, direction, and moisture are resolved as vortex streets. Hot moisture rising above a cup of tea or hot soup meets resistance in the cooler, calmer air and curls around but continues to rise, forming a series of growing eddies meshing along a vortex street.

A candle's flame wags back and forth, as a flapping flag does in the breeze, and for the same reasons. Both flame and flag move along the central zigzag current of the turbulence. Fly a kite with a long tail and observe the tail waggle as a vortex street, revealing the normally invisible rhythms of air.

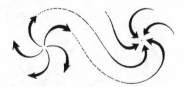

High pressure   Low pressure
(fair weather)   (storm)
(north of the equator)

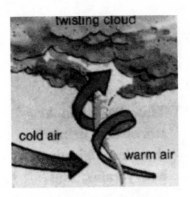

A tornado (in Spanish *toronado* signifies "thunderstorm") results from a clashing of contrasting temperatures and of winds. It is the fastest wind on earth approaching 300 mph, capable of destroying the instruments designed to measure it.

PLANTS AS LIVING TURBULENCE

Believe one who knows: you will find more in woods than in
    books. Trees and stones will teach you that which you can never
    learn from masters.

                    —*St. Bernard of Clairvaux (1090?–1153)*

The structures of growing plants often resemble patterns of water and air in motion. And why shouldn't they? Plants are mostly water, although they move much more slowly, so it is appropriate that they express the natural forms of flowing water. An important part of our own process of expanding vision is to see plants differently, not as static "things" but as dynamic *processes,* and to develop the ability to recognize plants, in whole and part, as living whirlpools, waves, mushroom rings, and vortex streets. We can learn to read nature's "hydroglyphics."

    Many plants, for example, obviously resemble upright vortex streets. The central stem corresponds with the zigzagging path flowing along the center of the vortex street. Each leaf, like a breaking wave, emerges alternately from the stem and curls around, its farthest tip turning back in search of the elusive eye. The architecture of a plant resembles the path of hot moisture rising into cooler, calmer, drier air. Its structure appears as a vortex street, giving us a glimpse of the invisible pattern of flowing energy, the lines

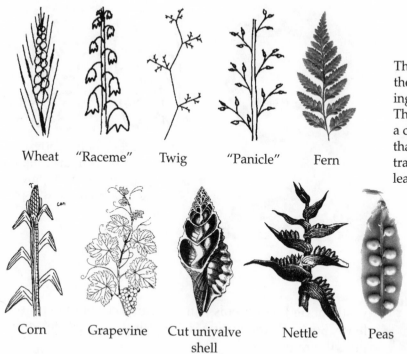

Wheat   "Raceme"   Twig   "Panicle"   Fern

Corn   Grapevine   Cut univalve shell   Nettle   Peas

The architecture of plants and their parts exhibits the rocking rhythm of a vortex street. This cadence also appears in a cut univalve shell as areas that alternate along the central column, resembling leaves growing along a vine.

of living force along which the plant's cells manifest like a bubbling froth to define its shape. Spirals occur in response to the world's resistance.

Plants and life in general are a kind of living turbulence. Living vortices, like liquid ones, have parts that model the whole. In larger plants and trees we can see how each branch is a miniature vortex street emerging from the central trunk. Each bough, branch, twig, stem, and leaf is a complete vortex street and part of greater vortex streets. When you're out among plants in a field, park, or forest, practice seeing vegetation as green fountains spraying up leaves in a slow whirl, like a slowly revolving lawn sprinkler. Watch leaves cascade on all sides of their slowly flowing green stems and vines. View leaves from up close and see that each one is like a miniature river delta, a stream of flowing life depositing living green silt upon the web of a growing pentagon. In moments of insight, as you practice this way of looking at vegetation, you may see all these eddies simultaneously as

Leaf structures often display vortex street rhythms.

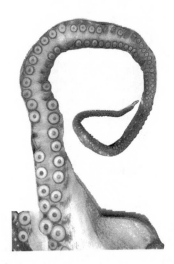

The pattern of the vortex street appears in the alternating placement of gripping suckers on an octopus's spiral tentacle, as well as the gripping tire treads.

The Bride of Frankenstein's wavy shock of white hair grows along the central path of a vortex street.

vastly nested whirlpools within whirlpools of living turbulence. Such self-replication is the hallmark of nature's spiral growth grammar.

RHYTHMS OF MANIFESTATION

From the contemplation of plants, men might be invited to Mathematical Enquirys.
           —*Nehemiah Grew (1641–1712, English plant physiologist)*

There's little doubt that the final years of the twentieth century are a time of rapid global, national, and individual transformation. Rising higher on the spiral of experience, a new world requires a transformed vision based on cooperation or partnership with nature, not exploitation. We'll have to see nature on her terms, as she is, abandoning the false images that have led to environmental disasters. Most of these delusions are centered around static pictures, nouns with distracting labels, snapshots, and pigeonholes, such as "greenhouse effect," "ozone depletion," "waste management," and "pollution control." Instead, we can learn to see nature as verbs, movies, *ongoing process,* and work with these to discover ways to prevent worsening conditions.

Insight into nature's geometric language will help us grasp her message. We can begin by learning to look at plants differently, any plants, from house plants and gardens, to the grocery's fruits and vegetables, to the great jungles and forests of the planet. We can train ourselves to view plants not as "things" but as *patterns of moving energy.*

Have you ever noticed how a plant grows over a period of days and weeks or how a flower opens, or the way new leaves unfurl on a vine? They spiral into existence, resembling whirlpools. Plants are not "things" but "energy events." It is only our slow nervous system that fixes surfaces and takes static "snapshots" of plants when the reality is continual change, slow process, energy in motion, a spiral path. Practice observing plants over time, in whole and in detail, and your vision will gradually shift to conform with life-facts rather than the incomplete pictures and clichés we file within ourselves. As you look at some living vegetation, use your imagination to visualize plants as graphs of energy in motion.

When you have seen plants this way you may want a deeper understanding of their rhythms.

A pinecone is helpful for this. Look carefully at one with an attitude of seeing anew and you'll notice that each bract—the brown, modified leaf that covers the seed ("pine nut")—lies upon a spiral chain of bracts winding from pole to pole of the pinecone. Parallel to this spiral chain are other spirals ("parastichies") that also wrap from pole to pole.

Each bract is also part of a spiral that spins in the opposite direction from that of the first spiral, at a different slant. Follow it around. Other spirals lie parallel to it. The point is that *each bract lies where two spirals cross.* Contemplate the pinecone as the physical manifestation of two opposite simultaneously spinning whirlpools. Where their lines of force cross, living matter manifests.

Once you begin to see in this way you may be surprised at how obvious the spiral architecture of plants has always been.

Living nature comes about by whirling into manifestation. At first, as invisible, living light, it thickens and swirls in energetic lines of force. Upon these lines the more dense substance of the ancient "world mother goddess" (the four phases of matter, or *mater*) precipitates, descending from light to gas to liquid and finally solid expression. The spirals of the pinecone are the material manifestation of the archetypal lines of dynamic energy. Where the lines cross, the bracts precipitate and become visible. Matter precipitates where positive and negative spiral rhythms synchronize.

All conifers—spruce, fir, redwood, giant sequoia, hemlock, and cypress—produce the modified flowers we call cones. The Swiss naturalist Charles Bonnet (1720–1794), known mainly for his studies of insect metamorphosis, collaborated in 1754 with the mathematician/artist G. L. Calandrini for the first modern mathematical study of branch and leaf arrangements, beginning with the pinecone. They coined the word *phyllotaxis* (Greek for "leaf arrangement") when they noticed the cone's definite structure of opposite-spinning spirals of bracts.

What they discovered was that counting the *number of parallel spirals* in each direction (not the numbers of bracts!) will inevitably yield consecutive numbers along the Fibonacci series, often five, eight, or thirteen. Their ratio of clockwise to counterclockwise spirals is an approximation of Φ, the golden mean. This ratio tells us that the cone has an inherent balance and harmony. If the numbers are low on the sequence the approximation to Φ will be crude. Higher numbers produce a ratio closer to the mathematical ideal. Each plant manifests the approximation that best suits its needs.

Countless examples exhibit this identical structure, although they display different numbers along the Fibonacci sequence.

Modern botanical thought on this subject is that the cone simply accumulates bracts, *mechanically* emerging wherever they fit. But deeper insight will show that the bracts are precipitating upon the invisible lines of force where opposite spirals cross. We are seeing the manifestation of the rhythm of an energy field. Just as the diminishing golden rectangle left a trail of whirling squares in its wake, the power of life leaves a trail of living matter in its wake as it whirls through the world. Plants are the visible clothing of the creating process.

Consider these examples. Better yet, observe living vegetation. The garden, greenhouse, and grocery are libraries of

I believe a leaf of grass is no less than the journeywork of the stars . . . and the running blackberry would adorn the parlors of heaven.

—Walt Whitman

Pineapples, raspberries, and blackberries are really many fertilized ovaries that have spirally accumulated to form one fruit.

The strawberry is an inside-out fruit with its seeds spiraling around the outside.

Corn, wheat, and many blossoms package their seed kernels in efficient whirlpools. Corn's kernels seem to form straight rows called *orthostichies* ("straight rows"), but take a closer look: the kernels spiral around and around the cob (eye).

The thorns, spikes, barbs, and needles of cacti and other plants wind around in spirals. These were once central veins in leaves, which diminished to conserve water in harsh environments, like pine needles.

The thorns of the rose wrap around in opposite spirals, revealing the subtle energy pattern of the archetypes of the creating process.

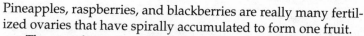

the Fibonacci rhythms of nature's creating process. Verify that each seed, fruit, and leaf manifests itself where two opposite-spinning lines of force cross. If you wish, count the numbers of parallel spirals in each direction and use a calculator to determine how close each is to representing the Φ ideal.

The architecture of upright plants shows that they put out leaves in spirals around the stem, like a spiral staircase

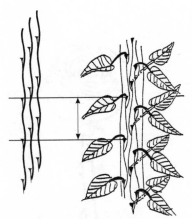

Within the plant's unfolded stem, its "leaf-trace" pattern feeds alternating leaves in a Fibonacci rhythm.

winding in either direction, clockwise and counterclockwise, at different slopes. The following illustration shows the leaf arrangements of some familiar plants and trees, listing their leaves/spirals ratio and characteristic angle between consecutive leaves, moving both clockwise and counterclockwise.

Within each plant's stem are veins, or "leaf traces," nourishing the leaves. Each vein feeds a chain of leaves along a particular pattern, passing those not in its pattern. The numbers of leaves between upper and lower leaves fed by the same vein comprise one "leaf cycle" reminding us of the wavelength between waves of the sea.

As your eye walks up and around the spiral staircase of leaves, you will discover that the number of leaves in one leaf cycle, that is, to the next leaf fed by that vein, is a Fibonacci number. Also, the number of spirals you turn to get to the next leaf cycle will *also* be a Fibonacci number. Each of the world's 350,000 known plant species can be characterized by the ratio of "leaves-to-spirals" in one leaf cycle. The higher the numbers are on the Fibonacci sequence and the more precise the ratio is to Φ, the more complex the plant (e.g., ferns two-to-one ratio is crude, sunflowers eighty-nine-to-fifty-five is finer). By multiplying the inverse

The ideal angle between consecutive leaves of a plant seen from above, the "sun's-eye view," provides minimum overlap.

"The leaf always turns its upper side towards the sky so that it may be better able to receive the dew over its whole surface; and these leaves are arranged on the plants in such a way that one covers another as little as possible. This alternation provides open spaces through which the sun and air may penetrate. The arrangement is such that drops from the first leaf fall on the fourth leaf in some cases and on the sixth in others."
—*Leonardo da Vinci (c. 1500)*

The green plant is the visible part of a spiraling energy field.

of this ratio by 360 degrees (to get a measurement of less than 360 degrees), you will discover the angle between measurement of consecutive leaves as they emerge from the stalk that is characteristic of the species. The ideal angle is approximately 137.5 degrees (= $1/\Phi \times 360$), or 222.5 degrees (its complement), measuring in the opposite direction. On a clock, the angle would be made by the hands at three minutes to four o'clock.

As Leonardo da Vinci noticed, there are supreme advantages to this spiraling arrangement. The more precise the Φ ratio is, the less overlap of leaves there is so that each leaf gets the most sunlight and the least shade. Plants seem to configure with the approximation to the ideal that best provides their needs for light, atmosphere, and balance. Look at plants from above, from the sun's-eye view, and notice that their spiral configuration allows you to see each leaf. The design also produces efficiently structured water collectors, solar water heaters and photovoltaic collectors, which are also roles of plants.

The golden spiral's geometric characteristic of self-similarity of part and whole guarantees that while the plant grows in size and weight it maintains the same balance, stability, and center of gravity.

A thing may endure in nature if it is duly proportioned to its necessity.

—Fra Luca Pacioli

Fibonacci fraction
of
$\dfrac{leaves}{spirals}$

| $\dfrac{2}{1}$ | $\dfrac{2}{1}$ | $\dfrac{3}{1}$ | $\dfrac{3}{2}$ |

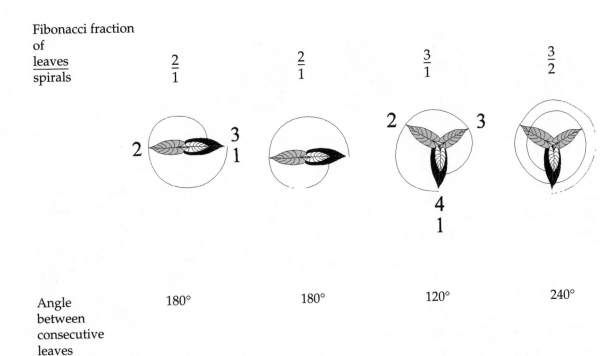

Angle
between
consecutive
leaves

| 180° | 180° | 120° | 240° |

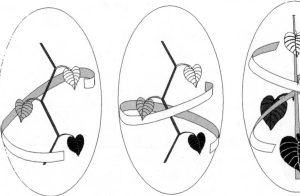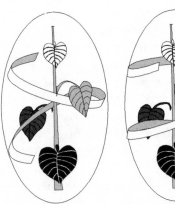

Leaf
arrangement:
Plant as
spiraling
field

Grasses, rice, corn, wheat,
sugar, bamboo, ferns, palm
fronds, lime, ivy, elm, bass-
wood, linden, ash, horse
chestnut, sycamore, maple,
dogwood

Beech, hazel, potato "eyes"

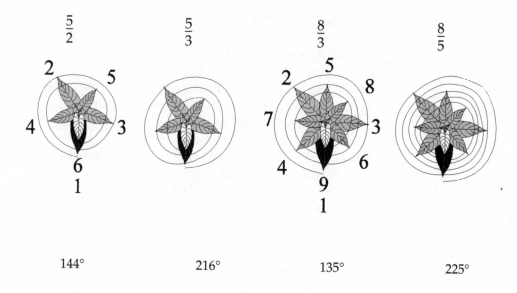

$$\frac{5}{2}$$ $$\frac{5}{3}$$ $$\frac{8}{3}$$ $$\frac{8}{5}$$

144°   216°   135°   225°

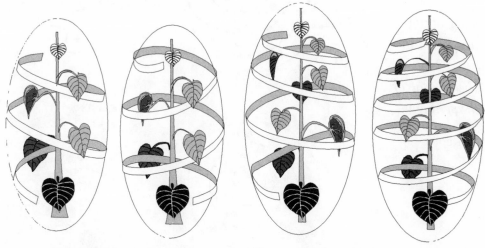

Apple, cherry, apricot, oak, cypress, poplar

Holly, pear, spruce, various beans

BUILD A PLANT REPLICA

With some skill it's possible to build models of the structural rhythms of many plants. You'll need stiff paper, scissors, a pencil, thumbtacks, and an upright rod mounted in a base (this could be a pencil in a ball of clay).

First, create or transfer the following spiral pattern onto firm paper or cardboard. Points along it are at ideal 137.5-degree intervals from each other. Note how each consecutive five points compose a whirling pentagonal pattern. Cut out the spiral and lift the eye into three dimensions, tacking it atop the rod so that the strip spirals around it. Tape the bottom to the base. Cut out paper leaves and glue or tape their

Where plants have five-fold patterns, a consideration of their souls is in place. For patterns of five appear in the regular solids, and so involve the ratio called the golden section, which results from a self-developing series that is an image of the faculty of propagation in plants. Thus the flower carries the authentic flag of this faculty, the pentagon.

—Johannes Kepler

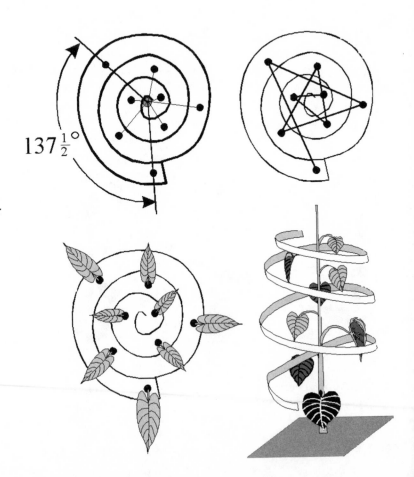

$137\tfrac{1}{2}°$

stems onto the rod behind each of the points. The number of leaves is a Fibonacci number, as is the number of spirals.

Any plant examples in the previous illustration can be replicated in three dimensions by replacing the points on the paper spiral with those of the plant's characteristic angle, using a protractor measuring out from the spiral's eye.

A needle and thread drawn through consecutive points along the spiral, when seen from the sun's-eye view, will create a pentagram star, the flag of life and excellence, showing that the principles of the Pentad are at the heart of living structures.

THE CALM "I" OF THE STORM

We've seen how the archetypal principles of number manifest themselves as forms around us. Circles and spheres, triangles, squares, pentagons, spirals, and the five Platonic volumes represent principles that shape the world. When we see stars and spirals we know that there is life and excellence, motion, balance, and transformation. But the universe is not like a room *filled* with geometric shapes. These forms are the shape *of* the universe, the shapes of space itself. What we see and touch are the visible expression of forces and principles we can train ourselves to understand. For the sake of solving our environmental and social problems, we can, and must, learn to see with new eyes.

The lesson of the spiral is that every "thing" is not a noun but a *process,* a dynamic "energy event." The world resembles a whirlpool of transformation with which we can cooperate for our benefit.

We can also use our experience to learn to see *ourselves* differently. A glance at the human body shows that in structure and function we carry the pentagonal flag of life and excellence. Like other living forms, our body's ideal proportions grow in accordance with the golden mean, the Φ relationship weaving balance, harmony, and beauty among its parts. This same accumulative proportion is found in the physical structure of living sea, land, and sky creatures and in the paths of atoms and galactic clusters. It is stirring to realize that each of the physical structures of the universe is

F or the world is not painted or adorned, but is from the beginning beautiful; and God has not made some beautiful things, but Beauty is the creator of the universe.

—Ralph Waldo Emerson

T he miracle is that the universe created a part of itself to study the rest of it, and that this part in studying itself finds the rest of the universe in its own natural inner realities.

—John C. Lilly

a repackaging of the others. We hold the proportions of the solar system in our hands, face, and whole body. One can fold an outline of the human body differently and create the proportions of a starfish and seashell, a rose, and the Milky Way galaxy. Perhaps the entire universe is a great, living, intelligent being we just don't recognize, in the same way that one cell in your body doesn't suspect the existence of the person reading this or even have the ability to comprehend your possibility.

By turning from the world outside ourselves to the subtle world within, we move from the symbolic to the sacred. In symbolic geometry the essential principle of the Pentad is life and "regeneration," expressed by the property of self-similarity. Inwardly, this characteristic indicates the possibility of *spiritual* regeneration, rebirth from the human to the divine as sought by every one of history's religions. Methods for attaining it differ, but curiously they share the spiral as a universal symbol of spiritual transformation, just as spirals indicate the transformation of forces in nature. In the Bible the Deity speaks to prophets from the same whirlwinds on which Native American initiates, like those in many other cultures, were carried to "heaven." The approach to heaven, not a location in the clouds but a symbol of regeneration to conscious identity with the Higher Self within, is never depicted as a straight line. Neither is the descent to hell, as Dante tells us from his downward spiral tour.

Craftspeople from ancient Egypt, Babylon, China, Africa, Europe, India, and Polynesia and great artists from Brueghel to Blake have symbolized the spiritual path as a spiral. All people seem to understand this intuitively. Even the makers of popular movies feel it. On her dream journey, Dorothy is transported first from Kansas to Munchkin Land on a spiraling "cyclone" and travels from there to Oz via the Yellow Brick Road, which begins, you will notice, at the eye of a golden spiral expanding ever outward to the Emerald City.

With its inherent harmony, the spiral is a metaphor for our inevitable transformation within. This inner spiral is not the physical spiral of our body, the embryo and ears, hair and heart, fists and fingerprints, but rather a psychic one,

Jacob's Ladder, a watercolor by William Blake, is depicted as a transforming spiral stretching between earth and heaven.

The Yellow Brick Road, symbol of transformation between Munchkin Land and Oz, begins as a golden spiral.

beckoning us as a symbol of *consciousness* and the way we live.

Our own inner spiral process whirls us into the mystery of the infinite within us. The spiral shows us that we are comprised of a continuum, not fragments, as we ordinarily appear to ourselves. A clue waits coiled in the spiral's principles of self-replication, self-accumulation, self-recurrence, and self-similarity. The recurring key word is "self." The message of the spiral is growth and the transformation of our Self.

At the center of our Self, deep within our consciousness, is a calm "I." Like the calm "eye" within a storm, our center is untouched by psychological turbulence. Peaceful, it observes all from the vantage of wisdom. Placid, it is unmoved by the turbulent weather of the surrounding psyche. When you're feeling connected with your center it seems very familiar. It feels like the Self you know best, like who and what you know your Self to be, calm in knowing

It is an ever-fixéd mark that looks on tempests and is never shaken.

—William Shakespeare

without thinking. To be centered is not the same as being "self-centered" or selfish. Instead, it is identity with the deep, divine power that motivates us.

Consider a spiraling top spinning "centered" on its calm eye. As soon as it tilts slightly, the top goes off center and its path widens, spiraling away from its center of stability, away from its center of gravity. The further from its center it moves, the more it wobbles. Leaving the center creates the trail of the spiral, an attempt to regain balance.

The center within us is the truth of our higher Self, the reality deepest within which each of us calls "I." Each time we lose sight of the "I" we spiral toward the periphery. We appear to leave the center because we feel a sense of separation somewhere in our lives. We lose our vantage point of the center when we artificially divide our Self and try to look at it, think about it, want it, grasp at it. When we leave our center, we are simply identifying with other possibilities of our Self, distracted by the turbulent realm of the four elements. And when we do, we miss the peace of the calm "I" and yearn to return to it, to transform our selves back to identity with it. Identity with our center is always possible, as a wobbling top can sometimes be helped to regain balance.

The further we wander from our calm center, the calm "I," the greater our experience of psychological turbulence. The motion of stepping away creates resistance, like moving an oar in water, leaving in its wake vortex streets of thought and emotion, desire and urge. Drifting from our center takes us away from the calm by creating resistance, resulting in the sometimes stormy turbulence of our psyche. Like water and air around us, fluid emotions and airy intellect within us moves in spirals. Like the golden spiral of nature, we grow by experiencing resistance and overcoming obstacles, adjusting our path as we go. We know we are off-center when we drift into identity with the clamor of sensations— the dramatic whirlpools of passion and fear and anger, dust devils of desire, swelling waves of emotion—and when we are kept awake at night by winds of thought.

It's a natural part of our growth process periodically to leave our center and come back to it, just as we must expand and contract to breathe. The creative pulse of the Dyad between center and turbulence is part of our growing

Without looking out of the window, one may see the way of heaven.

—Lao-tzu (c. 604–531 B.C., Chinese Taoist philosopher)

process. The rhythmic interplay between the impulse to divide ourselves from our center and the yearning that is expressed in our longing to return is found in the experience of constantly searching for and finding meaning in our lives. The interplay between these two forces manifests itself in the world as a curve. That's why the transformative path is characterized by a spiral and not another shape.

A spiral pulses all ways simultaneously and yet remains constant in its properties throughout change. By being aware of our inner motions and learning to observe ourselves, we can find the calm eye or calm "I" within the weather surrounding our center of awareness.

When we grow, we arrive at the same place where we were before, but we arrive more experienced, higher on the spiral.

There is nothing pleasurable except what is in harmony with the utmost depths of our divine nature.

—Heinrich Suso (1300?–1366, German mystic)

Structure-Function-Order

And God saw everything that he had made, and behold, it was very good. And it was evening and it was morning, the sixth day. Thus were finished the heavens and the earth, and all their host.

—Genesis 1:31, 2:1

Six days shalt thou labor, and do all thy work.

—Exodus 20:9

God has established nothing without geometric beauty which was not bound beforehand by some law of necessity.

—Johannes Kepler

All the pictures which science now draws of nature and which alone seem capable of according with observational fact are mathematical pictures . . . From the intrinsic evidence of his creation, the Great Architect of the Universe now begins to appear as a pure mathematician.

—Sir James H. Jeans (1877–1946, English physicist, astronomer,
and writer)

Six of one, half a dozen of the other.

—Folk saying

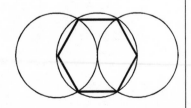

Birth of the Hexagon.

SIX APPEAL

We've now passed the halfway point on our journey to ten. "Six" is the fourth *actual* number according to the mathematical philosophers (who saw the birth of numbers at "three" and ending at "nine"). We could probably be understood around the world and throughout history because of the similar pronunciation of its name in Egyptian (*sas*), Assyrian (*sissa*), Sanskrit (*sas*), Hebrew (*sesh*), Arabic (*sitta*), Celtic Irish (*se*), Latin (*sex*), Italian (*sei*), German (*sechs*), Spanish (*seis*), French (*sees*), Russian (*sestj*), English (*six*), Danish (*seks*), and other languages. The archetype of sixness was called Hexad by the Greeks, in whose language the sibilant sound of "s" became the aspirated "h."

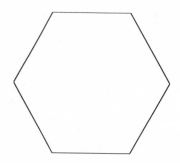

Whenever we encounter the Hexad in countless hexagonal (six-sided) natural phenomena and human designs, from beehives to faucet handles, we can be sure that certain principles are being called into play. By investigating the arithmetic properties of the number six and the geometric properties of the hexagon, we'll find that the Hexad is intimate with both Monad and Triad, the circle and triangle, unity and trinity, wholeness and balanced structure, applying their principles in an advanced way.

Hexagons contain a message that efficient *structure, function,* and *order* are occurring. But these three separate words represent a unity; the qualities they name are always integrated, never separate, and must exist simultaneously or not at all, like Borromean Rings. We'll refer to Hexad's principle as "structure-function-order" and "space-power-time." Scientists have recognized that separating space and time was incorrect, so they call it space-time. But there is a third factor, power, which must be included. Every whole event occurs at the intersection of these three aspects. Hexad represents their framework.

Consider any whole event or "thing" and recognize that it has a spatial structure, inherent power, and duration in time. What else is there?

Structure-function-order. Ponder the words separately, then feel them together as one concept, as a simultaneous, mutually reinforcing whole. Work into your "feel-knowing" that the *structure* of any event determines how it can func-

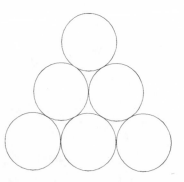

tion, that its functioning takes place in an orderly sequence, and that this order of unfolding determines what the structure must be. Sixfold phenomena serve as the visual representation of the Hexad's imprint of structure-function-order.

In surprising ways, multiples of six, particularly twelve, thirty-six, and sixty, also emerge from the principles of the Hexad as a "natural" framework in mathematics, nature, the symbolic arts, and everyday affairs. The dodecahedron's dozen faces, the traditional representation of quintessential "heaven," was associated with the twelve constellations of the ancient zodiac. The story of the process of the sun's journey through the cycle of the year told through the symbolism of the zodiac has endured for countless centuries as an archetypal tale of twelvefold progression on many levels.

The story of the starry skies was purposely mirrored in terrestrial affairs as the many twelve-part epic myths of solar heroes (as the twelve ordeals of Gilgamesh and the twelve labors of Hercules) and in a surprising number of religions (particularly those having twelve disciples around a central figure, modeling the zodiac around the sun). Corresponding twelvefold structure-function-order in architectural design and symbolism was essential to the astronomically aligned temples from Stonehenge to the Gothic cathedrals and Native American tepees. It's well known that whole societies from the Twelve Tribes of Israel to Solon's Athenian society and the Chinese and European courts were organized according to the duodecimal cosmic ideal. Even the proportions of everyday weights and measures were based on the natural twelveness evident in mathematics, which makes subdivisions and multiples easier (since twelve and sixty have so many divisors). This supreme structure-function-order of archetypal twelveness allowed ancient societies to move in harmony with the heavens and to cooperate with nature around them.

Embedded in these cosmic and terrestrial structures and stories were allegories known by every ancient initiate to disguise a teaching about our own inner archetypal structure-function-order. Knowledge of twelve-step processes was considered useful for the conscious development of a perfected individual, whom the Greeks called *Ho Nikon*, The Conqueror. He was represented as the Sun God surrounded by a pantheon of twelve spiritual attributes. Traditionally born during the winter darkness, dying and resurrecting in springtime, the sun's annual journey around the zodiacal belt symbolized the process of spiritual regeneration taught by the ancients all over the world.

The Hexad's principle of unified structure-function-order will allow us to incorporate in our universe structures from snowflakes to the starry zodiac. Have your outer and inner geometer's tools ready, and we will begin as near as we can get to the Hexad's archetypal birth.

A PERFECT SIX

The ancient mathematical philosophers began to understand the principles of any numerical archetype by first looking at its arithmetic and geometric properties. The unique properties of sixness will help us gain insight into the Hexad's principle of structure-function-order.

The Hexad has unique number properties. Six is the first of only two terms within the Dekad composed by the multiplication of two *different* factors (other than unity). It shares this attribute only with the completed Dekad (see chapter ten):

$$1 = 1 \times 1 \qquad \mathbf{6 = 2 \times 3}$$
$$2 = 1 \times 2 \qquad 7 = 1 \times 7$$
$$3 = 1 \times 3 \qquad 8 = 2 \times 2 \times 2$$
$$4 = 2 \times 2 \qquad 9 = 3 \times 3$$
$$5 = 1 \times 5 \qquad \mathbf{10 = 2 \times 5}$$

Because six is a doubling of three it partakes of the Triad's principle of balanced structure.

The Hexad is in many ways rooted in the Triad. The ancient mathematical philosophers, viewing numbers as collections of Monads arranged into geometric shapes, saw six as the third of four *triangular* numbers within the Dekad (1, 3, 6, 10).

They also noticed that six is both the sum *and* product of its divisors, the first three terms (1, 2, and 3). Thus, to the ancients, six represented the parents (1 and 2) of all numbers with their firstborn (3), thus making a completed whole.

They called it the Perfection of Parts, since six is the only

Triangular numbers grow as their bottom row increases by consecutive terms of the counting sequence.

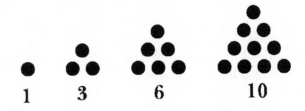

number that is both the sum and product of the same three integers. In fact, the divisors of six (1, 2, and 3) compose the *only* set of three integers wherein each number divides the sum of the other two.

Further intrigued by the interplay of two and three, and noticing that 2 + 3 = 5 and 2 × 3 = 6, the ancient mathematical philosophers saw in both the Pentad and Hexad different symbols of "marriage" since they're both formed by the interplay of a *female* term (two is considered even and female) and a male term (three is odd and considered male). And like biological parents, only five and six generate progeny like themselves; that is, self-multiplying powers of five always end in a five (5 × 5 = 25, 5 × 5 × 5 = 125, and so on). Likewise, the powers of six necessarily end in a six (6 × 6 = 36, 6 × 6 × 6 = 216, and so on). No other numbers (aside from unity, which generates and shares the properties of all numbers) have the ability to generate such models of themselves. The Pentad expresses this self-replication in its living forms. The Hexad expresses its self-similarity in self-reinforcing structure-function-order. Thus it was dubbed the Form of Form and the Unwearied Anvil by the Pythagoreans.

In all ways, then, the Hexad rests on a stable foundation whose parts are mutually supporting.

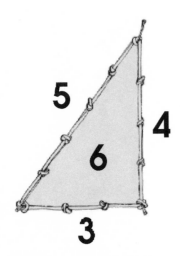

The Hexad is sometimes symbolized by the "Pythagorean triangle," or "3-4-5 right triangle," made by the ancient method using a twelve-knotted rope. It displays the sequence from one to six: one right angle, two unequal angles, sides of three, four, and five, enclosing an area of six square units.

THE BIRTH OF THE HEXAGON

Geometrically constructing a hexagon will show us how the Hexad's internal arithmetic structure manifests in universal design.

Since antiquity, various methods have been known for constructing hexagons, some of which are shown here. The hexagon is among the easiest forms to construct, emerging naturally from the circle, the Monad.

Method one: Birth through the *vesica piscis*. (1) First, construct a *vesica piscis*. (2) Use the straightedge to draw a line connecting the circles' centers, extending them to meet each circle's circumference. (3) Place the compass point at one end of the line and turn an identical circle to create a mirroring *vesica piscis*. (4) Reveal the hexagon by connecting the six points around the central circle, that is, the two centers and four points created where the circles cross.

Method two: "Walk" the radius around the circle. An even simpler method for constructing a hexagon, the one you probably learned in school, involves "walking" the compass open to a circle's radius around its circumference. (1) Center yourself and turn a circle. Keep the compass open to the size of the radius. (2) Put its point anywhere on the circumference and walk it around the circle, marking each point where the pencil crosses it. If you're careful, the sixth step will end precisely where the first step began. (3) Connect the points to form a hexagon, and then connect alternate points

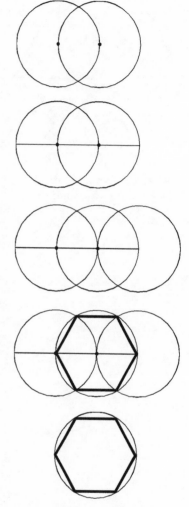

Method one: Birth through the *vesica piscis*.

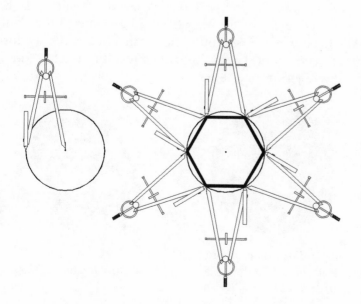

Method two: "Walking" a circle's radius around its circumference. The hexagon is an exteriorization of the circle's radius. Internally, it enfolds six Triads.

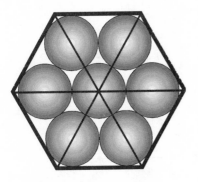

Method three: Six identical circles always surround one.

to form two interlaced equilateral triangles known as a "hexagram" star.

Every hexagon is a statement about the relationship of a circle's circumference to its radius. A hexagon is the circle's radius externalized, its circumference internalized. In other words, each one of a hexagon's sides is equal to its own radius and to the radius of the circle circumscribing it.

Method three: Six circles around one. It's a well-known geometric fact, although no one knows why it happens, that six identical circles fit precisely, or "kiss," around a central circle, each touching the center circle and two others. Prove this by arranging six similar coins, cylindrical glasses, or oranges around a central one.

To the ancients this arrangement represented the week, with six days of activity around a central Sabbath day of rest, just as the compass rests at the central still point.

Reproduce an Ancient Pattern

Use the six corners of a hexagon inscribed within a circle as the centers of six identical circles to produce a lovely decoration used from time immemorial upon pottery, tiles, wallpaper, and fabric. Construct a hexagon using method one. Extend the hexagonal pattern by continuing to place the compass point wherever two circles cross. This pattern has been used as a tool for self-observation. Gaze at it a while. Watch how various symmetries stand out and transform before your eyes. These shifts are a result of fluctuations of our own nervous system.

Tile pattern (Pompeii).

Islamic tile and ornament pattern based on the hexagon, called a "true lover's knot."

Lincoln Cathedral East Window tracery, c. 1260.

Construct an Escher-type Design

The Dutch artist M. C. Escher (1898–1972) applied the ancient knowledge of hexagonal tiling patterns with images from nature and fantasy to create intriguing designs. You can, too. Just follow these directions as you look at the illustration:

With a light pencil and straightedge, subdivide a hexagonal grid into a field of triangles. (1) Start at any single triangle and draw a curve between two of its points. You can use the curve in this illustration to create whirling birds, or you can make up your own curve and be surprised at what you create. (2) "Pivot" the curve at one corner of the triangle and draw the exact same curve along the sides of six connecting triangles that make a hexagon. (3) On the original triangle's remaining side draw a different curve, one whose halves reflect each other in mirror symmetry around the *center* of this side. (4) Repeat this curve around the outside of the hexagon. (5) Now take a moment to design the bird, using simple lines for its spine, wings, and tail feathers. (6) Apply this design to the six birds of the hexagon, alternating dark and light, or with colors, to distinguish them. (7) Repeat this process in adjacent hexagons, noting how

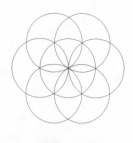

Pennsylvania Dutch "hex" sign.

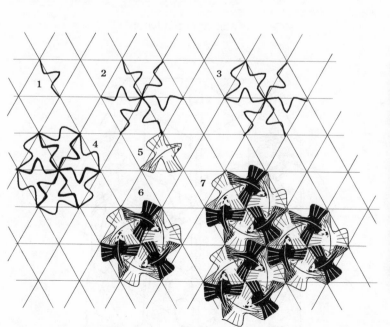

A Hindu *yantra*, or meditative diagram.

three hexagons mesh, with three light (or dark) tails meeting at a common point, to fill the sky completely with interlocking birds, leaving no gaps.

With practice and by seeing the results of different initial curves, you can get quite good at designing interesting tiling patterns.

Construct a hexagon. You'll need two identical strips of paper, each as wide as half the hexagon. Lay each strip on the hexagon and fold it (as shown) upon the hexagonal guide. Carefully join the two folded strips so they link. This knot is known to sailors for its ability to link two ropes together with great strength.

Simply stated in the language of natural design, the first principle of the Hexad is efficient, economical structure.

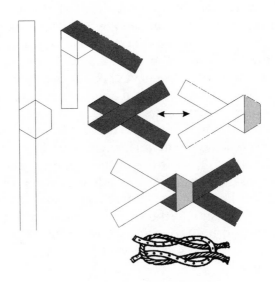

Tie a Hexagonal Knot.

The two ends of the Nile symbolically unite ancient Egypt with a hexagonal knot. The whole picture was designed in a hexagon whose diameter is the height of the pole. Connecting its corners and midpoints shows the figures aligning with, and partitioned by, the hexagonal geometry.

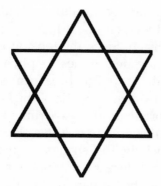

The hexagram star is an ancient symbol appearing in worldwide religion and myth. It is sometimes called the Seal of Solomon (which Moslem legend recounts him using to capture magically *djinn*, or "genie," nature spirits). As the revered symbol of the Jewish people, it is called the Shield of David. And to Hindus, it is the Mark of Vishnu. In some traditions, the hexagram star is the great seal of initiates, signifying rising aspiration from below met by the descent of grace from above.

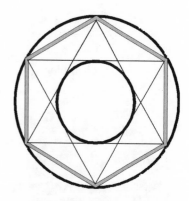

LISTENING TO STRUCTURE

The hexagon holds within it three distances: the length of its side (equal to its own radius), its diagonal (twice the radius), and the distance between alternate corners (root three [= 1.732. . .] times the radius). We can see in the illustration that the three lengths form the sides of a right triangle having sides one, two, and root three within the hexagon. These are the numbers associated with the circle, line, and triangle. A hexagon makes visible the relationships between the archetypal principles of Monad, Dyad, and Triad.

The relationships among these numbers are abstract but can be approached in another way. We can make mathematical relationships sensible by *hearing* them. Number relationships can be made audible through music. Try this: construct a large hexagon on a flat piece of wood and hammer nails partly in at the three corners of the inner right triangle as shown. Tightly wrap a single guitar string, piano wire, or

A curious fact is that a circle constructed within the inner hexagon of a hexagram star has a circumference exactly half that of the large outer circle. Verify this by measuring both with a piece of string.

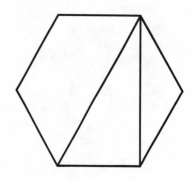

Nuts and bolts

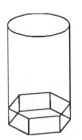

Glass tumbler

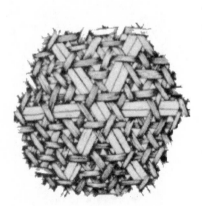

Center of a woven basket

Faucet handle

nylon fishing line tightly around the nails, keeping equal tension between them, and tie it tightly. Then simply pluck the lengths and listen. Since the tension is equal, only the *length* of each string determines its pitch. The shorter the string the higher the tone. The three form a related sequence of tones, a harmony characteristic of the hexagon.

What is important in this experiment is not each tone itself but their *relationships to each other.* We can hear the relationships by plucking the strings sequentially and then simultaneously in different combinations. They compose a harmony, based on the "square root of three" (= 1.732. . . to 1), the ratio within an equilateral triangle, the Triad, the archetype of balance.

HUMAN HEXAGONS

Hexagons appear endlessly in human inventions, providing greatest structure-function-order, that is, maximum efficiency of materials, labor, and time, by using straight lines to approximate the efficient space of the Monad, a circle. Its relationship with the Triad gives this form the great strength of three-corner joints. Designers need not know anything about the mathematics of the hexagon to discover that it works. Some hexagons stand alone, and some tessellate, or tile, in repeating patterns.

The single hexagon works in situations requiring strength and stability, using the fewest materials and straight lines. We know its effectiveness when we grip a faucet handle or tighten "hex nuts." A hexagon provides stability at the base of a tumbler and in the six central ribs of a woven basket.

Bicycle racers know that bicycle wheels are most efficient in terms of having the least weight and wind resistance yet the greatest strength if the number of spokes is a multiple of six, usually thirty-six.

Any design that requires a circle's efficiency but is limited to straight lines, like the light-limiting diaphragm of a camera lens, will somehow make use of the hexagon. Circular umbrellas, caps, and parachutes make the best use of material with the least weight when they consist of six sections, or a multiple of six.

Human ingenuity may make various inventions . . . but it will never devise any inventions more beautiful, nor more simple, nor more to the purpose than Nature does; because in her inventions nothing is wanting and nothing is superfluous.

—Leonardo da Vinci

Many familiar items rely on hexagonal structure.

HEXAGONAL PACKAGING

When hexagons arrange in repeating patterns they extend the pattern of "six-around-one." In this way the circles, and the hexagons formed by connecting their centers, are said to tessellate. Circles, cylinders, and spheres packed in a hexagonal pattern fill space more efficiently than in a square pattern, getting more into the same amount of space. Engineers and designers study this principle, but most people learn it at the market by observing the way fruit stacks in six-around-one tessellations.

But when circles tessellate they leave curved triangular gaps between them. Wasted space is little problem when stacking apples, but more efficient space-filling requires that the circles be converted to hexagons, leaving no gaps at all. Hexagonal tessellation extends the circle's principle of "equality in all directions" in a balanced sharing of space, materials, time, and energy. It creates three-corner 120-degree joints, which provide the triangle's strength and structural excellence to the package (see chapter three).

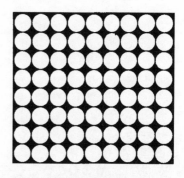

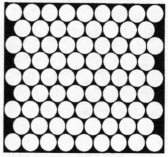

Hexagonal packaging incorporates more circles in the same area than square packing.

Hexagonal tiling leaves no gaps. The efficient pattern of six-around-one is so common that we hardly notice its familiar appearance.

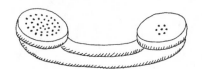

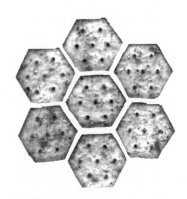

Computer chip circuit

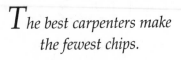

The best carpenters make the fewest chips.

—German proverb

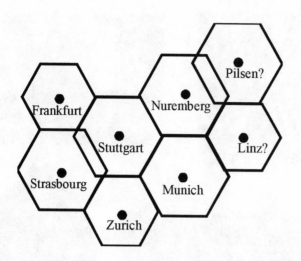

Central marketplaces. Studies find that older cities around the world seem purposefully arranged in a hexagonal lattice, a pattern that economizes market centers, decreases competition, minimizes transportation distances, and maximizes administrative control of outlying areas.

Once you begin to notice hexagonal tessellations you'll see them endlessly.

Before you read further, construct (or doodle) some hexagonal tessellations using circles or hexagons. Get a feel for their pattern of expansiveness, efficiency, and cohesive strength in close-packing.

HEXAGONS, NATURALLY

While pentagonal symmetry is only found in living structures, the Hexad manifests naturally in both living and non-living forms. All natural forms are forces made visible. When we encounter six-sided, six-angled, or six-pointed crystals, plants, and animals we know that the Hexad is underlying an efficient structure-function-order. Its three-corner 120-degree joints and close-packing arrangement ensure an efficiency of materials, time, energy, and strength.

The tessellating hexagonal pattern is common in the realm of molecules and crystals. Molecules typically occupy two thirds to three fourths of the crystal; the rest is empty space. A close-packed arrangement uses minimum space and best balances the attractive and repulsive forces to achieve the minimal total expenditure of energy. A knowledge of archetypal geometric patterns is the basis for the

Observe how a froth of bubbles in the kitchen sink eventually settles into six-around-one arrangements as the molecules strive to balance forces with minimum surface tension. The flexible reptile-skin pattern expresses the ideal energy pattern of forces in balance. The facets of a fly's eye are arranged in close-packed hexagonal arrangement.

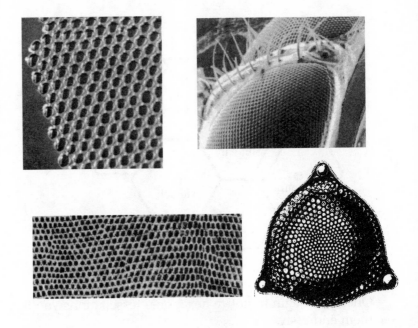

modern science of crystal engineering, the fabrication of molecular crystals with unusual optical, electronic, and magnetic properties.

We can't easily see arrangements of atoms and molecules, but we can see their pattern on a larger scale in a froth of bubbles, in snowflakes, and in quartz crystals.

Since nature traffics in efficiency, hexagonal tessellation recurs endlessly in nature. The molecular structure of wood cellulose is a hexagonal web, seen on a larger scale as water-net algae (hydrodietyon). The star coral configures a hexagonal skeleton, and tube coral grow by leaving their cylindrical skeletons packed together in hexagonal groups. Notice the same flexible hexagonal tiling in the arrangement of scales on a fish or in the skin of a reptile, like the chain-mail net of knights' armor. A tortoiseshell, with its hexagonal plates, also takes the shape of minimum energy and maximum strength.

In the human body individual cells join in a close-packed hexagonal pattern to act in concert. The so-called striped muscle that occurs throughout our bodies is hexagonally packed and available for voluntary movement, such as when you move your eyes along the page to read these words. The fishing net pattern recurs in human lung alveoli, maximizing the passageways for oxygen–carbon dioxide transfer in each of our breaths.

Single hexagons are commonly encountered in living configurations. A slice of carrot and the top of a green pepper reveal hexagonal structure, as do microscopic radiolaria, diatoms, viruses, insects, and some sponges.

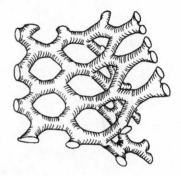

Human lung alveoli form a hexagonal net.

A carrot slice and green pepper top, like the top of a tomato and other fruits and vegetables, reveal their hexagons.

*D*oth not nature itself
teach you?

—I Corinthians 11:14

All true insects have six legs. This water strider walks upon the water surface's molecular tension as if on a net, lifting alternate legs in stable tetrahedral fashion.

Who carved the nucleus, before it fell, into six horns of ice?

—Johannes Kepler

The appearance of crystalline snowflakes is why scientists consider water a mineral. As the temperature drops, H_2O molecules vibrate more slowly, slow enough for electric charges within each molecule to attract other molecules and tighten into a hexagonal, close-packed arrangement. More molecules build upon the seed pattern to become beautiful snowflakes, blanketing the world with sixfold symmetry.

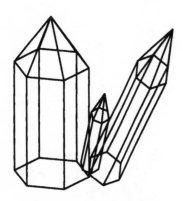

Close-packed silicon atoms arrange in a hexagonal pattern like ice and grow as a six-sided quartz crystal. Due to its electromechanical, or piezoelectric, properties, pressure on quartz yields electricity, while electric current put into quartz from a battery produces regular mechanical pulses, hence, the "tick" of the "quartz watch." Even in this most modern sense the Hexad is associated with order and timekeeping.

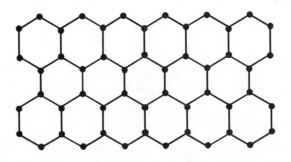

The graphite in pencils is made not of lead but carbon, the same element that comprises diamonds. But diamonds have tetrahedral architecture, making them nearly indestructible, while graphite crystals grow in hexagonal sheets that slide off the pencil as we write.

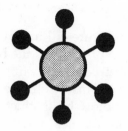

Uranium hexafluoride
(UF_6)

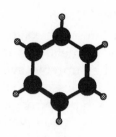

Vitamin C

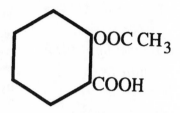

Aspirin

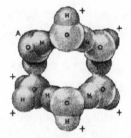

Six water molecules form
the core of each snowflake.

Cyclo-octa-deca-nona-ene,
an aromatic fragrance

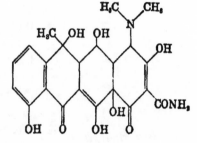

Terramycin

The discovery of the benzene ring (C_6H_6), the basic structure of organic chemistry, came to the German chemist Friedrich Kekule after he dreamed of the *ouraboros,* the circular snake biting its own tail. The Hexad's principles configure wherever efficient structure-function-order is required. Molecular chemists and biologists find them in innumerable familiar substances including steroids (cortisone and the male and female hormones testosterone and estrogen), cholesterol, benzene, TNT, vitamins C and D, uranium hexafluoride (nuclear fuel), aspirin , sugar, the antibiotic Terramycin, and pencil graphite.

The atomic structure of cyclo-octa-deca-nona-ene, an aromatic fragrance, reminds us of Islamic patterns and beehive structure.

[T he existence of pentagons and hexagons] doth neatly declare how nature Geometrizeth and observeth order in all things.

—Sir Thomas Browne

*F*or so work the honey-
bees,
Creatures that by a rule in
nature teach
The act of order to a
peopled kingdom.

—William Shakespeare

*T*here being, then, three
figures which of themselves
can fill up space around a
point, viz . . . the triangle,
square and the hexagon,
the bees have wisely
selected for their structure
that which contains the
most angles, suspecting
indeed that it could hold
more honey than the other
two.

—Pappus (c. 300 A.D., Greek
mathematician in Alexandria,
Egypt, whose writings
stimulated seventh-century
geometric studies)

Honey Space

The most famous hexagonal architecture is built by bees, wasps, and hornets. The ancient Egyptians took the honeycomb as the supreme symbol of structure-function-order and applied its lessons to their society. Insects, tutored only by the archetypal principles they know by instinct, build hexagonal nests and hives with three-corner joints to minimize labor and maximize interior space and strength.

Honeybees, one of the very few domesticated insects (silkworms are another), build their honeycomb in alternating hexagonal tiers like a wax crystal. After the bees have eaten large quantities of the sweet substance, wax oozes from special pockets in their bodies. They chew this wax until it is soft and work it into the familiar hexagonal pattern. Due to the space-filling strength of the hexagonal shape and structure of three-corner 120-degree joints, the least wax is used to hold the most honey: a mere one and one-half ounces of wax holds four pounds of honey.

Hornets and wasps, on the other hand, also build hexagonal nests but of mud or a paperlike substance whose strength also relies on geometry. But unlike the bees' honeycombs, which have a horizontal entrance, wasps' nests hang vertically and are entered from below, while hornets prefer to enter their hives from above.

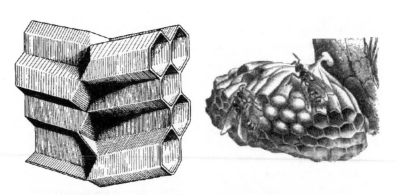

Different views into the structure-function-order of a beehive. A wasp's nest entered from below.

DODEKAD—TWELVE—THE FRAMEWORK NUMBER

Twelve, called Dodekad (in Greek literally "two ten"), emerges from the principles of the Hexad. Although not included within the first ten terms of the Pythagorean Dekad, it relates intimately with each of them, even the "virgin" number seven (see chapter seven).

As double six, the Dodekad extends and refines the Hexad's structure-function-order comprising whole events. The wonder of twelve is that it has so many divisors (one, two, three, four, and six), making it the supreme number appointed at the archetypal level of mathematics as the natural framework of arithmetic and geometry. Its properties have made it the number of symbolic measurement and sacred structure since prehistory.

Construct the Dodekagon

Among the many methods for constructing a twelve-pointed or twelve-sided structure, three stand out for their ease and antiquity. They are based on the fact that twelve is divided by three, four, and six, the only three numbers whose shapes (triangle, square, hexagon) tessellate, covering a flat surface without leaving gaps. Use your tools to try each of these methods. Be free to explore each construction further.

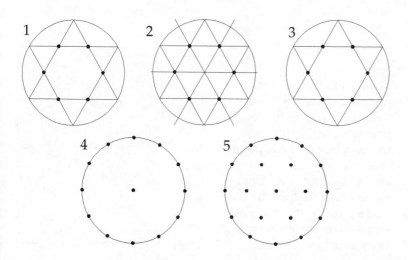

First method: (1) Construct a hexagram star within a circle. (2) Draw straight lines across the circle's center through each of the six crossings, extending them to reach the circle. (3–4) With the original six points we now have twelve equally spaced points around the circle, like a clock face. Discover the star's patterns by doodling, connecting various points with straight lines or sewn thread. Use the new crossings as centers to turn more circles.

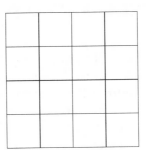
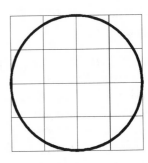

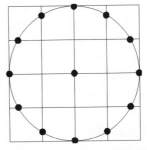

Geometry existed before the creation.

—Johannes Kepler

Second method: Construct a square and divide its sides into four equal parts, drawing a grid as shown. Inscribe a circle within the square, a Monad within a Tetrad. The circle crosses the grid at exactly twelve equally spaced points, like the numbers on a clock. Twelve-part wholeness emerges when the transcendental Monad reconciles with the rational Tetrad. Twelve represents the crystallized expression of divinity.

Third method: Construct a square and inscribe a circle within it. Keeping the compass open to the circle's radius, place its metal point at each of the four corners and midpoint of each side. Make an arc (or turn the full circles) from each point. The arcs cross the original circle at twelve equispaced points.

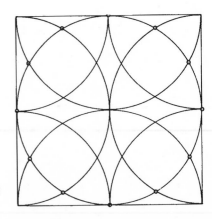

Twelve and Thirteen

Twelve is related to thirteen, the notorious but misunderstood number of superstition. While twelve is a solar number, thirteen is lunar, as the twelve months of a solar year contain nearly thirteen lunations. The relation is more obvious when we look geometrically. Each of the methods for constructing a dodecagon results in twelve equally spaced points around a thirteenth one at the center. Twelve points occur at the crossings of a hexagram star around the thirteenth at their center. We'll look at this geometry as it evolved in many symbolic expressions through the centuries.

The pattern of twelve and thirteen continues in three dimensions. Exactly twelve equal spheres pack precisely to enclose a central thirteenth sphere.

If you have some clay and toothpicks, make thirteen equal-size balls and join them to see how twelve fit around or "kiss" one at the center. Each sphere touches the center and four others around it in an omnidirectional closest-packing. If you remove the central sphere, the remaining twelve collapse slightly to create an icosahedron, the twenty-sided Platonic Volume.

Known and studied in ancient times, this volume appears naturally in galena and other crystals. Crystallographers know the shape as a "cuboctahedron." You can make one from a cube of clay by slicing tetrahedra off its eight corners, from the midpoints of its edges.

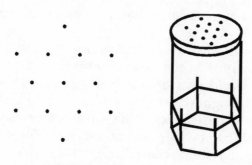

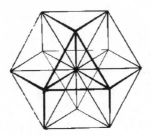

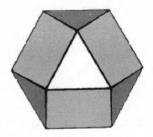

Cuboctahedron crystal is a graph of balanced forces, or "vector equilibrium," in twelve directions out from the center.

The cuboctahedron (or "dymaxion" as it was called by Buckminster Fuller) is an omnidirectional graph of forces in balance. Its twelve lines radiate from the center as a "vector equilibrium," or truss, balancing three-dimensional forces most effectively.

Any way we look at it, in two or three dimensions, twelve-around-one composes a wholeness, unity, perfection, a universal module of structure-function-order.

ZODIACAL SOCIETIES

A father has twelve children. Each has thirty daughters, one side white, the other side black, and though immortal, they all die. Who is the father?

—*Cleobulus*

(Answer: the year)

The patterns by which the cosmic creating process constructs the universe are consistent on all levels. Wherever efficient structure-function-order is required the Hexad appears.

The prehistoric transformation from nomadic tribes to settled agricultural societies occurred at about the same time around the world, soon after the ice age ended about 12,000 B.C. People brought with them their myths and understandings, which included tales of the sun and night sky, as well as a surprising knowledge of mathematics. The daily cycle of the rising and setting constellations that had served as the nomad's clock as it progressed through the year indicated periods of soil preparation, planting, care, and harvesting and thereby determines the structure of the agricultural calendar.

Contrasted with our modern civilization with its technological wonders, prehistoric societies were primitive. But there is evidence that their thinking was much more sophisticated than we might imagine. They often consciously modeled their communities on the twelvefold pattern of constellations they perceived in the night sky. To the tribe, the whole nation symbolized the twelve-constellated zodiacal wheel that they replicated on the landscape, often reshaping it with mounds and ditches. They divided themselves into twelve provinces, each corresponding to a zodiacal month. Coins unearthed by archaeologists often show the zodiacal symbolism of the city at which they are found. Each province was divided into thirty clans like the days of the month, which were further subdivided into thirty houses. The Greek leader Solon devised his famous Athenian constitution this way. Each province was also allotted one-twelfth of the national myth, and each choir sang music based on one note of the twelve-note chromatic musical scale.

To the mathematical philosophers of ancient Greece, "heaven" beyond, which encompassed the four elements, was symbolized by the fifth Platonic volume, the twelve-

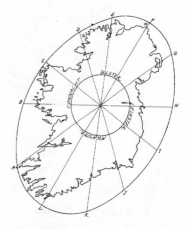

The twelve-part division of ancient Ireland with its center at Tara, the High King's celestial court, was designed as a microcosm of the zodiac.

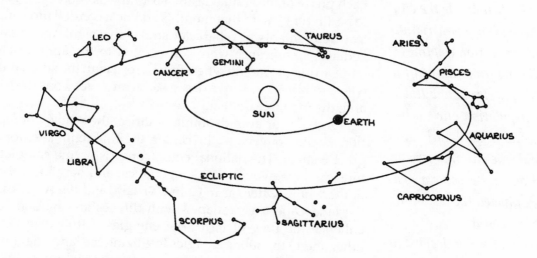

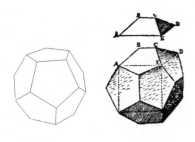

The legislator's first job is to locate the city as precisely as possible in the center of the country . . . Next he must divide the country into twelve sections. But first he ought to reserve a sacred area for Hestia, Zeus and Athene (calling it the "acropolis"), and enclose its boundaries; he will then divide the city and the whole country into twelve sections by lines radiating from this central point . . . He should divide the population into twelve sections, . . . [and] . . . divide the city into twelve sections in the same way as they divided the rest of the country . . . Now that we've decided to divide the citizens into

faced dodecahedron. Plutarch, in discussing this form, divides each pentagonal face into thirty triangles, making 360 triangles in all. Thus, the dodecahedron represents the thirty clans within each of twelve provinces. At the same time it represents the full circle of 360 degrees and the cycle of the year of 360 days (plus five intercalary festival days).

The practical benefit of this division of a nation is that with the number twelve's many divisors, the society could be easily subdivided to perform different functions. But, more important, the philosopher-leaders envisioned that by mirroring the night sky on earth they were participating in the harmonies of the cosmic structure-function-order.

As John Michell and Christine Rhone show in their book *Twelve-Tribe Nations*, there is abundant historical information to confirm that the ancients gave great importance to a twelvefold order of society. Many of these "twelve-tribe nations" are revealed by their twelve-gated cities, myths of twelve disciples around a central figure, and cosmologically ordered royal courts.

The populace of these twelve-tribe nations participated in an annual course of pilgrimages, festivals, and a circle of perpetual choirs following the seasonal flow of terrestrial fertility around the zodiacal wheel of their landscape. At the full moon, when the earth's magnetic energy peaked, a major festival was held at the holy site—temple, holy spring, well, or cavern—in the province appropriate for that month. Each phase of the annual festival round was choreographed to be "in tune" with that month's zodiacal constellation and featured symbols and characteristics of that sign. Animals of zodiacal significance were consecrated to the appropriate city. During the month when Taurus rose with the sun, a bull was prominent; when Aries rose, a ram was consecrated and, during Pisces, a fish.

Once a year at midsummer sunrise, the whole population would convene at the center, the thirteenth site, for a great festival. The national choir, comprising the complete musical scale, would lead song, dance, and harmonization of the whole. Often, as in China, Britain, and the Americas, the landscape was reshaped with ditches and mounds to enhance the flow of terrestrial energies. Attuned to each other and to the subtle energies fertilizing the land, the choir would amplify the natural fusion of terrestrial forces with

A clerk, astronomer, and recorder (*Psalter of Blanche of Castille*) study celestial geometry to calculate festival dates. The picture itself was composed with the same twelve-fold symmetry the three men study in the heavens.

twelve sections, we should try to realize (after all, it's clear enough) the enormous number of divisors the subdivisions of each section have, and reflect how these in turn can be subdivided and subdivided again . . . This is the mathematical framework which will yield you your brotherhoods, local administrative units, villages, your military companies and marching-columns, as well as units of coinage, liquid and dry measures, and weights. The law must regulate all these details so that the proper proportions and correspondences are observed.

—Plato

celestial energies and literally enchant the land and the society. In this way the society was ruled through music. Although ancient knowledge of sound vibrations is a long-lost science, it is slowly being rediscovered through such practices as voice toning and experiments in cymatics (the process through which sound vibrations establish geometric patterns in matter).

Countless civilizations are known to have regulated themselves by the twelvefold pattern. According to the early Greek historian Strabo, Egyptian society was originally divided into twelve *nomes,* or provinces, each ruled by a king, thus externalizing the myth of Osiris and his twelve retainers.

Greek legend and myth are full of reference to the duodecimal pattern. Foremost was the Olympian pantheon of twelve gods and goddesses presided over by Zeus at the center. Apollo, Greek sun god (sun king), was crowned with twelve powers. Odysseus sailed with twelve ships commanded by twelve followers. The ordeals surmounted by Hercules express the nature and characteristics of each of the twelve phases of any whole. The legendary Cadmus is said to have founded the center of Thebes by driving a stake through a dragon (thereby harnessing the fertilizing earth energies) and sowing its twelve "teeth," which became the founding families. In this sort of foundation myth we can see the transition from lunar nomadic life to solar settled societies.

At the hub of its twelvefold society, the sacred national Greek center at Delphi had a stone *omphalos,* or "world navel." The renowned Delphic Oracle sat on a golden tripod in a cavern over a fissure and spoke in hexameter verse later used by Longfellow and other poets.

In the Roman foundation myth twelve vultures appear to Romulus indicating the exact site (on seven hills) for the city. Referring to the terrestrial zodiac, the Latin word for "city" (*urbs*) is related to the word "round" (*orbis*). Roman law was displayed in public written on twelve bronze tablets.

Religions and governments mirroring archetypal twelveness abounded throughout the world's early settled societies. We've heard about the twelve tribes of Israel, the Vedic Kings surrounded by twelve nobles (as was the Norse god Odin by his twelve counselors), the Dalai Lama's Round Council of twelve great Namshans, the Irish High Court at Tara, and Charlemagne's mystical court. In the twelve-part epic legend of King Arthur and his Round Table of twelve knights, King Arthur represents the sun and the "round table" is, of course, the twelve-constellated zodiacal belt.

Native America was once a twelve-tribe nation, each tribe consisting of twelve clans. The traditional sweat lodge and tepee structures are still ritually constructed to resemble the cosmic pattern of a calendar resting on twelve poles of the months, with certain objects and symbols of each month appropriately placed around the periphery.

Less well known is the existence of the twelve-around-

The Native American tepee is ritually structured as a calendar.

one pattern in the government of the United States, the founders of which were Freemasons. Seeking to establish a utopian society in the ancient tradition, they gave the government a classic twelvefold structure-function-order. In every way possible they incorporated zodiacal symbolism. George Washington kept a circle of twelve generals around him to fight for the thirteen colonies. The symbolism of thirteen stars and thirteen stripes on their flag, thirteen letters in their motto, "E Pluribus Unum" ("From many, One"), and thirteen buttons on the sailor's uniform accented their intent. The number of colonies was not accidental but a hidden reference to an ancient teaching.

The country's founders put the two-sided Great Seal of the United States on every dollar bill. In chapter three we saw the seal's Egyptian pyramid with thirteen courses of stone capped with the triangular eye of divinity. The other side of the seal depicts the great eagle as the Egyptian hawk Horus. Above the eagle's head is a glory of light and clouds enclosing pentagonal stars explicitly displaying the twelvefold canon. When the glory's geometry is magnified and applied to the seal as a whole, the underlying plan on which it was composed becomes plain.

DISCOVER CLASSIC HEXAGONAL COMPOSITION IN ART

Try this with tracing paper over the Great Eagle Seal on a dollar bill: put a point at the center of the circular seal at the eagle's chest just above the shield. Put the point of the compass there, and open it to the circle's radius. Turn a circle. Confirm that it matches the seal's circle. Starting with the compass point at the top, inscribe a hexagon within the circle by walking the compass around its circumference. Connect the six points to inscribe a hexagram star. If your pencil is sharp, inscribe hexagram stars within each inner hexagon. The construction on the tracing paper will reveal the obviously planned composition of the seal underneath as shown in the illustration.

The identical hexagonal structure underlies a large silver plate called *David Trying on Saul's Armor* from a similarly

Eagle side of the U.S.
Great Seal

David Trying on Saul's Armor

Aegeus before the Oracle of Delphi

Bella Coola wooden sun mask

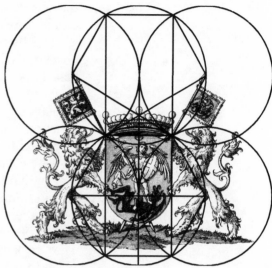

British Royal Seal

designed set of four plates from Byzantine Constantinople (610–630). Although it was designed over one thousand years earlier than the U.S. Great Seal, it employs the same underlying framework for the proportionate placement of figures. Note how the "omega" near the top corresponds with the glory of clouds above the eagle.

Symbols and art of many societies that centered around a sun king were often designed upon this hexagonal lattice, a symbol of divine order on earth, the pattern of solar societies since prehistory. The twelve dark squares around the painting on a plate from Delphi, a twelve-structured society, gives us a hint as to the geometry underlying its design. From Native American sun masks to British Royal Seals, hexagonal and twelve-structured geometry symbolizes both the form of society and the perceived cosmic structure with which it is meant to harmonize. Two hexagons (twelve points and crossings) have been marked around the Delphic painting and Native American sun mask for you to connect in order to discover their geometric guidelines.

BREATHING WITH THE COSMOS

In their desire to live in harmony with nature and the perceived structure-function-order of the heavens, the ancients applied the twelvefold cosmic canon to every detail of their terrestrial affairs. Not only was the overall pattern of society based upon the duodecimal frame, but multiples of six (particularly 12, 24, 36, 60, 360, 600) were used to coordinate measures of structure and space (distance, area, volume), of function (weights, coinage, administration) and of order (timekeeping).

The canon even integrated language into its workings. Ancient Greek, like Egyptian, Hebrew, Arabic, Coptic, Syriac, and Sanskrit (literally "well-constructed") used letters as numbers. By adding the values of its letters, important words of myth and cosmology revealed their significance. The Greek name for the Egyptian river Nile, ΝΕΙΛΟΣ, transposes into 50+5+10+30+70+200. Their sum, 365, refers to the

We've inherited our "English system of measure" from the ancient twelvefold canon. Although the modern metric system derives from the aftermath of the French Revolution as a measure of the earth (the meter is 1/10-millionth the distance from the North Pole to the equator through Paris), it is not part of a comprehensive system integrating celestial dimensions and cycles with human life.

Structure (Space)

12 inches = 1 foot
36 inches = 1 yard
6 feet = 1 fathom (from the Saxon *faetm* for
    "embrace" of extended arms)
660 feet = 1 furlong ("furrow length," the distance
    plowed in one acre, stretched out in a straight line)
360 degrees = measure of a circle
60 minutes = 1 degree of arc
60 seconds = 1 minute of arc

Function (weights and other measures)

12 = 1 dozen
12 dozen = 1 gross
(originally) 12 ounces = 1 pound

Order (Time)

60 seconds = 1 minute	30 days = 1 month
60 minutes = 1 hour	12 months = 1 year
24 hours = 1 day	360 (+5) days = 1 year

Nile's 365-day cycle of rising, overflowing, depositing fertile silt, and receding. The Egyptian number word for sixty was "hen." The Greek word *henad* ("unity") has a letter value of sixty. And the Greek word for universal structure-function-order, ΚΟΣΜΟΣ (cosmos), has a value of 600, emphasizing the importance of six and its multiples in cosmic design.

The modern world has inherited remnants of the ancient system of measure from Chaldea through Babylon, where the captive Jews learned it and passed it on to Rome and Europe. In the United States it is known as the "English" system.

Over the centuries, the ancient relationships have changed somewhat. The pound weight was originally equal to twelve "ounces" (from the Latin *uncia* signifying "one-twelfth") and only later became sixteen ounces. From the court of Charlemagne the British derived their monetary system of twelve pence to the shilling and twelve shillings to the pound sterling, used until the adoption of the metric system in 1971. The twelve-person jury derives from this canon, as does the analog clock face. The metric system makes calculations easy but dissociates us from a frame linked with the cosmos.

The system of duodecimal measure is so widespread one would think various civilizations had collaborated on it. But independent of each other, China, Babylon, Egypt, Greece, and India all divided the day/night cycle into twelve two-hour periods (which the Greeks called "Babylonian hours") and sixty twenty-four-minute parts, further subdivided into sixty 24-second parts. A Chinese clepsydra (water clock) was made of six water pots in descending sequence, dripping from one to the next. A Hindu hemispherical copper bowl called a *kapala* had a hole in the bottom; when floated in a basin of water, its hole allowed in just enough water to slowly fill and sink it sixty times in twenty-four hours.

In ancient cultures timekeeping meant more than calculating festival dates and hours for prayer. It was a way that humans could see themselves integrated within the cosmic scheme. They accomplished this by making the basic unit of measure equal to the duration of one human breath (= 4 seconds). Structure-function-order were united and measured in terms of the human breath as distance-breath-duration.

I am not one who was born in the possession of knowledge; I am one who is fond of antiquity, and earnest in seeking it there.

—Confucius (551–479 B.C., Chinese philosopher)

This ancient Chinese clock, like a mandala, was also a candle. As it burned along its labyrinthine path, it marked off the minutes and hours of the day.

Enlarging by scales of six and sixty they found their place among the dimensions of the earth, sun, moon, and stars.

The ancient Hindu system of measure shows how this system works.

1 prana (breath)	= 4 seconds
6 prana	= one vinadi = 24 seconds
60 prana	= 10 vinadi = 240 seconds = 4 minutes = time necessary for earth to rotate 1 degree on its axis
60 vinadi	= one nadi = 360 breaths = 1440 seconds = 24 minutes = time necessary for earth to rotate 10 degrees
60 nadi	= one sidereal day/night = 21,600 breaths = 86,400 seconds = 1440 minutes = 24 hours = time necessary for earth to rotate 360 degrees

Observe due measure, for right timing is in all things the most important factor.

—Hesiod (c. 800 B.C., Greek poet)

The people of ancient cultures saw themselves and their culture breathing in unison with the turning earth and stars. The numbers reveal the cosmic structure they perceived. The number of seconds in one day (86,400 = 12 x 12 x 660) is exactly one-tenth the diameter of the sun (864,000 = 12 x 12 x 6,000) in miles. The numerical values of the Greek letters of the name of Pythagoras, named after the Pythian oracle who predicted his birth and greatness, adds up to 864, a certain reference to his solar stature.

And the day's 21,600 (= 6 x 60 x 60) breaths is exactly ten times the moon's diameter of 2,160 miles. Astronomers also know that the sun travels at 21,600 miles per hour through the galaxy as a satellite around its unknown center. This number is also one tenth of the mean distance *between* the earth and moon (= 216,000 miles = 60 x 60 x 60), which is 100 times its own radius. The moon's diameter in miles resonates with the 2,160-year measure of each Platonic Month, or Zodiacal Age. Twelve such ages complete a Great Year, one cycle of the Precession of the Equinoxes, as the earth slowly wobbles on its axis in 25,920 (= 6 x 12 x 360) years. The unit of the mile, which allows this hidden structure of the cosmos to emerge, is much older than its attribution to 1,000 (in Latin *mille*) paces of the Roman Legion implies. The mile was defined so that about 25,920 of them equal the circumference of the earth around its equator, in resonance with the time of a precession.

There's little sense in arguing whether the ancients knew the dimensions and cycles of the earth, moon, sun, and stars. We know so little about the seers of old and the deep antiquity from which these units of measure derived that our conclusions, colored by today's very different vision of our place in the cosmos, can only be speculative. But by noting that the number values of the Greek words for "year" (*etos* = 575) and "breath" (*pneuma* = 576) are just a single unit apart, we can get an indication of the basis of ancient measure.

The consistent predominance of multiples of the numbers 6, 12, 60, 360, and 600 indicate that the manifest cosmos is an approximation of its mathematical ideal. We can marvel at the ancient systems of measure, constructed as they were with a sophisticated knowledge of the archetypal framework numbers and incorporating everything from a single human breath to the size of the sun and the 25,920-year cycle of the Precession of the Equinoxes.

These number correspondences are not coincidental but recur wherever the cosmos is perceived as an integrated whole. The ancient duodecimal units of measure were carefully chosen and coordinated by seers to integrate human structure-function-order as distance-breath-duration with celestial rhythms and dimensions. When appropriate units are chosen the celestial pattern shows through.

While we measure with rulers and weigh ourselves on scales, most people don't realize the vast and intelligent system we've inherited, mirroring the cosmos within which we live and function. These tools are fossils of a long-forgotten way of looking at ourselves as integrated within universal relationships. This knowledge of ourselves as part of a harmonious whole is now dimmed to everyday awareness, but it can be part of a great vision to be reclaimed.

THE RECONCILING TEMPLE

In ancient times design conveyed more than just style. Strict rules based on archetypal symbolism existed for the design of everything from kitchen utensils and furniture to temples and

A tradition which has been credited by many learned men over the centuries is that the ancients encoded their knowledge of the world in the dimensions of their sacred monuments.

—John Michell

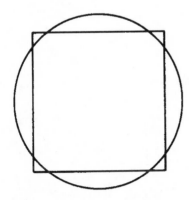

Squared circle. A square and circle of nearly equal perimeter and circumference. Perfect equality is impossible to achieve with the geomter's tools.

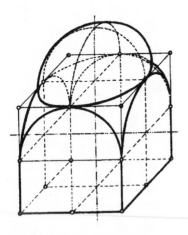

A Byzantine cupola reconciles the "heavenly" sphere with the "earthly" cube, forming a three-dimensional version of the squared circle.

monuments. Temples and society were both meant to mediate and harmonize the heavens with the affairs of people. Through number and geometric symbolism cosmology was built into the measures and proportions of sacred architecture. From prehistoric megaliths to Gothic cathedrals, temple design is embedded with profound and surprising knowledge of relationships among the dimensions and cycles of the earth, moon, sun, planets, and stars. In accordance with the overall twelvefold structure, the temple was designed as a timekeeping device, a measuring tool, both a clock and a calendar mirroring the patterns of heaven on earth.

Harmony of heaven and earth was symbolized in ancient times by a geometric puzzle posed by the Delphic Oracle and known as "squaring the circle." The challenge was to construct a square whose area or perimeter was equal to the area or circumference of a given circle. This puzzle has obsessed scores of people over the centuries without solution because it cannot be solved with the three tools of the geometer, except approximately. It requires an impossibility: the absolute reconciliation between the transcendental circle, whose immeasurable circumference teases us with a never-ending decimal, and the square's rational sides. The ideal and the actual, the archetype and its representation, can never be identical.

A "squared circle" is the ultimate symbol of reconciliation of the heavens with the earth. Thus, it is approximated in the symbolic geometry and measures regulating the architecture of many temples, whose role is such a reconciliation.

Although a squared circle cannot be constructed precisely with the geometer's tools, two *vesicas pisces* constructed at right angles to each other yield a good approximation, which was known in ancient times. The square formed by the four innermost crossings has a perimeter nearly equal to the circumference of a circle made within the *vesicas pisces*. This symbol of geometric harmony became the core of the temple's design.

Square the Circle

Do this ancient construction for squaring a circle, creating a circle whose circumference approximates the perimeter of a square. Ordinarily, it would be done ritually on the earth at the site of the proposed temple.

1. Construct two *vesicas pisces* at right angles to each other.

2. Connect their four innermost crossings to make a square.

3. Inscribe a circle within the two *vesicas pisces* through the square. This is the "squared circle" whose circumference approximately equals the square's perimeter. Confirm it by measuring each figure with a piece of string.

4. Inscribe a circle *within* the square, representing the earth.

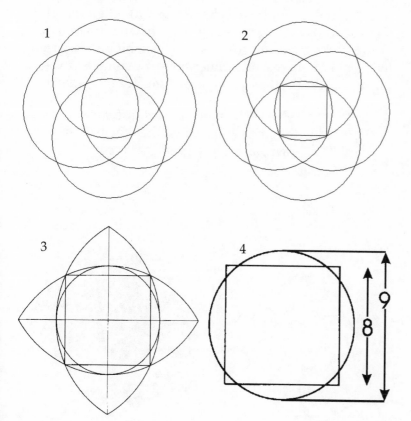

*M*easures, to plot the sky . . .
I use a ruler, thus, until at length
The circle has been squared, and in the midst
A market place is set, from which the streets
Are drawn, to radiate as from a star,
The beams of which, itself a circle, shine
Straight forth to every point.

—Aristophanes (c. 450–c. 388 B.C., Greek playwright)

5. *Atop* the square construct a small square whose center is on the outer ring. Inscribe a circle within it, representing the moon.

The geometry of the squared circle discloses the inner harmony of the earth-moon relationship. The inner circle of earth within the square and small circle atop it (the moon) show the relative diameters of the earth ($7{,}920 = 8 \times 9 \times 10 \times 11$ miles) and moon ($2{,}160$ miles). On each side of the moon's square is a three-four-five right triangle, easily formed with the use of a twelve-knotted cord. In this way the ancient seers studied the archetypal patterns of geometry and encoded celestial dimensions within temple design.

Music was also incorporated into this scheme. The relationship between the diameters of the large circle of the moon's path and the circle of the earth come close to the ratio of nine to eight, the basic interval of the "tone" on which the musical scale is built (see chapter seven). The ratio of the areas of the circles is close to Φ, the golden ratio of natural growth. The dynamic cosmos was worshipped as visible, living music, its celestial dimensions revealing an expanding musical scale, the Pythagorean "music of the spheres."

6. When twelve such "moon circles" are constructed within the band around the "circle of the earth" as shown, and a hexagram star is con-

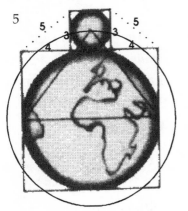

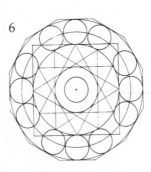

structed within the band, we see a more complete scheme rich in number relationships. For example, the combined radii of the earth and moon equal 5,040 (= $1 \times 2 \times 3 \times 4 \times 5 \times 6 \times 7 = 7 \times 8 \times 9 \times 10$) miles, the number of the "ideal population" of Plato's utopian twelve-tribe nation of Magnesia described in his *Laws*. Plato disguised these numbers of the cosmic ideal throughout his writings to hint at the numerical and celestial scheme he wasn't allowed to discuss overtly. The celestial design canon contained in the Delphic riddle of the squared circle served as the pattern for twelve-part temples, myths, measures, and societies throughout the world over many centuries.

Stonehenge and Gothic cathedrals are examples from the long tradition of temple design that encode such knowledge in their geometric proportions.

Discover the Geometry of Stonehenge and Chartres

A squared circle has been constructed on these images of Stonehenge and the altar at Chartres. Explore their geometric scheme using your compass open to the radius of the inner circle. Put the point of the compass at the top of that

The North Rose Window of the Cathedral of Notre Dame at Chartres displays the twelve-part structure of the ancient canon.

Stonehenge was constructed first of wood around 2700 B.C., later of stone, built in stages ending around 1500 B.C.

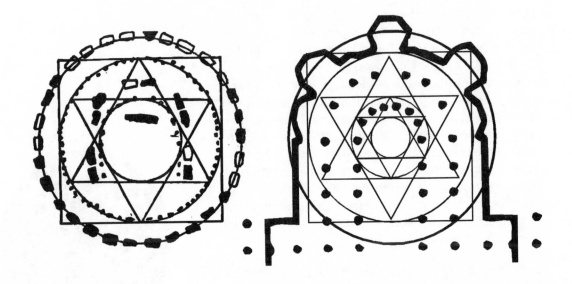

Stones placed along Stonehenge's outer circle, representing the moon's path, were used to predict the moon's risings, settings, and eclipses, as well as solar rhythms. The altars of Chartres and other cathedrals were designed with the squared circle and hexagonal fusion geometry of Stonehenge, as seen in its Rose Window, which repeats the twelve-part geometry of its altar.

*D*o *not seek to follow the footsteps of the men of old; seek what they sought.*

—Matsuo Basho (1644–1694, Japanese poet)

circle and walk it around to mark the six points of a hexagon. Connect the points to draw a hexagram star within the hexagon, as seen earlier in the geometry of the Great Seal of the United States. The geometry of Stonehenge's stones and Chartres' altar area are aligned with these interlaced triangles. Symbolically and functionally both are temples that served to fuse and harmonize the energies of heaven and earth, sky and soil, the entire cosmos and a single human breath.

THE SUN-BORN SELF

Modern science knows that there are twelve chemical steps, like twelve notes of the chromatic music scale, in every spiraling turn of the DNA helix, determining our genetic makeup. The ancient mathematical philosophers, not having the tools of modern science but seeing how the number twelve relates to all numbers, recognized that the functions of all numbers group themselves around twelve. With the unchanging properties of number as a model, they codified this as a natural frame of the universal structure-function-order. While the twelvefold cosmic canon served as model for all aspects of society, there was another level to its symbolic architecture, myths, and measures. A personal inner significance was given to initiates who had achieved a degree of psychological purification and were ready for more serious growth and spiritual transformation. Since they were strictly confined to an oral tradition, the details of what they were taught is not known. But it's no secret that many of the mystery schools of ancient cultures sought a spiritual rebirth within the individual using the zodiac as a teaching tool in an original twelve-step self-help program.

From myths, legends, and art we know that the zodiac was used as a sublime allegory of initiation. The twelve cryptic zodiacal constellations were seen nightly by everyone; for the initiate, they were translated into tools of self-knowledge. The zodiac described the inner constitution of humans, the structure-function-order of ourselves, mirroring that of the universe. According to the ancients, the archetypal principles revealed by arithmetic and geometry, represented by the zodiac, seen manifest in the designs of nature, technology, art, the clock and calendar, temple architecture, myths, mea-

*Wouldst read the story
of the self-born King?
First learn the splendid
language of the sun,
The speech of the stars, the
moon's coy whispering,
The music of the planets,
and of one,
Our Mother Earth,
crooning her cradle-song
To her uncounted babes,
who, when they gain
The soul's full stature, to
the heavens belong . . .*

—James Morgan Pryse

Mithras, personification of the Higher Self, is represented as the archetype of the year. Surrounded by the zodiac, his spiritual powers, he pips his energy field and bursts from his Orphic Egg, born into identity with his Higher Self. The Greek letters of the name of Mithras, Μειθρασ, add up to 365, reinforcing his symbolism as the Monad of the year.

The Circle of the Zodiac as the Quest for Self. Fifteenth-century personification and zootyping of the drama of the energies and principles of the sky by Barthelemy l'Anglais (*Book on the Properties of Things*).

sures, and whole societies have correspondence within each of us. The initiate was taught to see the world inside-out, to see twelve-part wholes as *our* reflection.

This is the inner significance of the dodecahedron, the fifth Platonic volume with twelve five-sided faces, symbol of heaven. Five is a symbol of humanness. Each face represents a different matrix of ordeals and opportunities before us. The whole dodecahedron represents "heaven," our spiritual completion.

NORTH—Midnight—Winter solstice—Darkness.
   Sun rises furthest south, travels lowest in the sky.

I Seek My Higher Self Through What I Use—Capricorn

I Seek My Higher Self Through Humanity—Aquarius

I Seek My Higher Self Therefore I Am—Sagittarius

I Seek My Higher Self and I Don't Seek My Higher Self—Pisces

I Seek My Higher Self Through What I Desire—Scorpio

I Seek My Higher Self—Aries

I Seek My Higher Self Through What I Balance—Libra

I Seek My Higher Self Through What I Have—Taurus

I Seek My Higher Self Through What I Learn—Virgo

WEST—Sunset—Autumn equinox—Declining light. Night and day balance.

EAST—Sunrise—Spring equinox—Birth of light. Day and night balance.

I Seek My Higher Self Through What I Think—Gemini

I Seek My Higher Self Through What I Learn—Virgo

I Seek My Higher Self Through What I Feel—Cancer

SOUTH—Noon—Summer solstice—Greatest light.
   Sun rises furthest north, soars highest in the sky.

Although different myths were conceived for different cultures, each in its own way depicted exploits of the unified Cosmic Being, the ideal of every age. The twelve attributes represented by the zodiac were assembled into an ideal teacher, who guided the spiritual welfare of humanity during that 2,160-year period. In their cryptic language, solar allegories described a plan of spiritual birth within ourselves, of purification through twelve types of ordeal.

The twelve signs of the zodiac, twelve disciples, twelve tribes, and pantheon of twelve gods and goddesses represent twelve archetypal principles within everyone. These are our own twelve phases of transformation, the soul's journey home with Odysseus through twelve ordeals, the stages of expanding growth necessary in order to become complete and enlightened. Each individual comprises all twelve aspects of the whole, and we will eventually find within ourselves the principles represented by the symbols.

The twelve-part national myth of twelve-tribe nations is a tale about both the visible sun and the spiritual light within ourselves. A savior is born during the darkest days of the winter solstice; his light grows and dims through the twelve signs of the zodiac. As the sun goes through the hours of the day and months of the year, we experience a correlated process of development. This process can be seen in a great deal of Egyptian art, which is often composed on the grid of a calendar.

The myths and legends of Apollo, Heracles, Osiris, Ra, Gilgamesh, Odysseus, Jesus, Mohammed, Arthur, the twelve harvesters in the Egyptian Field of Amenti, twelve carpenters, potters, weavers, fishermen, twelve rowers in the ship with Ra, twelve stones chosen by Joshua, and twelve facets of the ancient solar deities all symbolize our innate faculties of spiritual intelligence, twelve grades of being, twelve stages of unfoldment. These twelve phases of transformation are twelve ways we seek identity with our Higher Self. This drama is repeated on all levels from the subatomic to the supergalactic as the great and mysterious power of consciousness in the universe develops through a round of twelve stages of identity at each level.

The ancient Chinese system of acupuncture recognizes our body's twelve-part symmetry as having twelve "meridians," or subtle nerve channels, around a central thirteenth

Astrology is a science in itself and contains an illuminating body of knowledge. It taught me many things, and I am greatly indebted to it. Geophysical evidence reveals the power of the stars and the planets in relation to the terrestrial. In turn, astrology reinforces this power to some extent. This is why astrology is like a life-giving elixir for mankind.

—Albert Einstein

I loved you, so I drew these tides of men into my hands and wrote my will across the sky in stars.

—T. E. Lawrence ("of Arabia," 1888–1935, British archaeologist, soldier, and writer)

And when, monks, in these four noble truths my due knowledge and insight with the three sections and twelve divisions was well purified, then monks . . . I had attained the highest complete enlightenment.

—Buddha

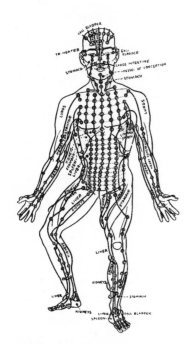

This Chinese acupuncture map of the body, a replica of the cosmos, recognizes twelve subtle nerve channels around a central thirteenth, an archetype of the zodiac and twelve spheres surrounding one.

meridian through which *ch'i,* or creative energy, flows. This is the same fertilizing *ch'i* that courses seasonally through the body of the earth and that twelve-tribe societies sought to amplify through their temples and festivals. Although dowsers are known worldwide, the only ancient fertilizing system remaining intact is the Chinese science of *feng-shui* (pronounced "foong-shway," literally "wind and water"), the geomancy of orienting and siting temples, tombs, homes, and societies for the purpose of "enchanting" and fertilizing the land, family, and nation. *Feng-shui* brought harmony to the environment in ancient China as acupuncture harmonizes the energies of the body.

The initiate was instructed to know him- or herself as a dynamic energy system like that of the earth. Consciously directing energies to flow in constructive channels was done for the human's body and the earth's. Before one could reach enlightened awareness of his or her identity with the whole cosmic creating process, it was necessary to tame and purify both the energies of the land and the initiate's turbulent inner "private world."

In an advanced stage of meditation the head's twelve cranial nerves are energized to liberate the individual as an enlightened being, a solar conqueror, a "self-rolling wheel," world savior, avatar, messiah of the age, like a central sphere surrounded by twelve others, symbolizing twelve attributes, or powers.

Today the ancient system is in fragments; each of the sciences, arts, and religions contains remnants of it. Because we insist on a literal interpretation of past wisdom, it has been distorted and obscured. We make undue efforts to see "things" and people outside ourselves as having power over us. Instead of discovering the principles of twelve-around-one within our deeper self, we relegate it only to a pattern of holes in a salt shaker and a game of Chinese checkers.

Yet we cannot and should not expect to rediscover the full body of ancient wisdom by studying dusty monuments and myths full of idioms and subtle references understood only by those who lived at the time. The perennial wisdom requires each individual and age to discover it anew in external mathematics, expressing it in ways and symbols suitable for those times and cultures.

CHAPTER SEVEN

HEPTAD

Enchanting Virgin

But the seventh day is the Sabbath of the Lord thy God; in it thou shalt not do any work . . .

—Genesis 20:10–11

Wisdom hath builded her house, she hath hewn out her seven pillars.

—Proverbs 9:1

If you keep a thing seven years, you are sure to find a use for it.
—Sir Walter Scott (1771–1832, Scottish poet, novelist, historian, and biographer)

Measure seven times before you cut.

—Carpenters' saying

He found the vast Thought with seven heads that is born of the Truth.

—Maha Upanishad

A lie has seven endings.

—Swahili proverb

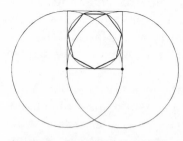

A regular heptagon cannot be constructed with the geometer's three tools and so is not born like other shapes through the *vesica piscis*. But an approximate heptagon is possible to construct.

ABOMINATIONS AGES-OF-MAN
ALTARS ANGELS ANIMALS
ASSISTANTS BEARS BIRDS
BLESSINGS BOWLS BREATHS
BROTHERS CANDLES CAVES
CELEBRATIONS CHAMBERS
CHAPTERS CHIEFS CHILDREN
CHOIRS CIRCLINGS CITIES
CLEANSINGS CLEAN-BEASTS
COINS COMPANIONS
CONSTELLATIONS CONTINENTS
COVENANTS COWS CROWNS
CURSES DAUGHTERS DEFENDERS
DEGREES-OF-WISDOM DEVILS
DOGS DOORS ELDERS EUNUCHS
EYES FISHES FRAGRANCES
GABLES GATES GENERATIONS
GIANTS HEADS HEAVENS HONEST-
MEN HORNS ISLANDS KINGS LAKES
LAMPS LEADERS LETTERS
LIBATIONS LOAVES MATRIARCHS
METALS MOUNTAINS MYSTERIES
NATIONS OATHS OVERSEERS PATHS
PATRIARCHS PETALS PILLARS
PLAGUES POOLS PRIESTS
PRIESTESSES PRINCES
PRINCESSES PROSTRATIONS
QUEENS RAYS REPETITIONS
RIDGES RIVERS RUNGS
SACRIFICES SAMURAI SEAS
SINGERS SINS SISTERS SLEEPERS
SNAKES SNEEZES SONS
STAIRCASES STARS STEPS STONES
STREAMS SWANS TEMPLES
THUNDERS TREES TROUBLES
TRUMPETS VEILS VENGEANCES
VIALS VIRGINS VIRTUES VOICES
VOWELS VOYAGES WARRIORS
WATCHERS WISE-MEN WISE-WOMEN
WOES WONDERS YEARS

*H*ail! *Great Mother, not hath been uncovered thy birth.*

—Egyptian Hymn to Neith

FOLK SEVENS

Seven is perhaps the most venerated number of the Dekad, the number par excellence of the ancient world. Its name has been similarly pronounced around the world: in Sanskrit *saptan*, Egyptian *sefek*, Latin *septem*, Arabic *sabun*, Hebrew *sheva*, Spanish *siete*, and so on. The Greeks, whose word for seven was *hepta*, nicknamed it *"sebo,"* or "veneration." Seven has always predominated in mythology, mysticism, magic, ceremony, and superstition (both lucky and unlucky). Its appeal endures to this day in religion, nursery rhymes, and folk sayings, from the reverent "seven angels before the throne" to the colloquial "seven-year itch." Our feel for sevenness is a measure of our sensitivity to the archetypal principles of the Heptad.

The Bible, literary source of Western culture, contains thousands of references to sevens, heptarchies, and septenaries, the Greek and Roman terms for sevennesses. Seven lions were in the den with Daniel, and seven locks of Samson's hair were cut. Joshua's seven trumpeters circled Jericho seven times. There were seven voyages of Sinbad and seven Wonders of the Ancient World. Seven perambulations are made around the Kaaba at Mecca. Animals considered clean were admitted to Noah's Ark by sevens, unclean animals by twos. The seven "liberal arts" of medieval education were so named because they represented seven paths of learning intended to *liberate* us from a mundane life. Snow White and the Seven Dwarfs is an old tale containing hidden wisdom within its symbolism. You probably know many more references. What is the source of our fascination with seven? What are the principles of the Heptad for which we accord it such special attention?

Seven is on everyone's lips at one time or another. When asked to think of a "lucky" number, you're more than likely to choose seven or three. Lottery games often disallow the number 777 to prevent excessive payout due to its popularity. Think of something familiar that comes in sevens. Some people recall that the opposite sides of honest dice add to seven, but most think of the seven days of the week.

Unlike the Monads of the day, month, and year, which are measures of the motions of the earth, moon and sun, the ancient measure of a "week" (from the Gothic *wiko*, signifying "sequence to which we come," which derived from the Egyptian *uak*, the word for "festival") is not based on any celestial cycle, although it is conveniently close to being a fourth part of the moon's monthly cycle of 29.5206 days (one "moonth"). While the concept of weekdays and weekend grew out of the Industrial Revolution, there is something in us that moves with the seven-day rhythm. Seven somehow creates a familiar cycle.

In many religions every seventh day is a Sabbath (from the Hebrew *shabbat*, "to cease from labor"), a sacred day distinct from the six-part structure-function-order of the worldly weekdays. The weekday is both sacred and profane, light and dark, God-made and man-made, timelessness within time. But on the Sabbath there is only holy and holy; there is only light, as nothing else is attended to. Duality is suffused into the realm of the eternal. The Sabbath is the virgin bride of the week.

Traditionally every seventh year was a "year of release," when a field was allowed to rest from planting, when debts were forgotten and slaves were set free. Today, the law sets seven years as the duration of the statute of limitations for some crimes. Folk wisdom refers to life proceeding in seven-year periods. Consider your own life. Does it make sense to see your development as stages between the ages of zero–seven–fourteen–twenty-one–twenty-eight–thirty-five–forty-two–and so on?

A group of seven comprises a complete unit, a whole event. But a group of seven is different from other wholes we've encountered, particularly the Monad, Triad, and Hexad. The Heptad expresses a complete event having a beginning, middle, and end through seven stages, which keep repeating. Seven represents a complete yet *ongoing* process, a periodic rhythm of internal relationships. All configured efforts are led in seven stages to perfection. To look deeper into the principles of the Heptad, we have to look at the unique arithmetic and geometry of the number seven.

MATHEMATICS OF THE VIRGIN

The number seven occupies a critical place within the Dekad, where it acts as both a link and a chasm. As a link between the first six and last three terms, $1 \times 2 \times 3 \times 4 \times 5 \times$

All the world's a stage
And all the men and
women merely players;
They have their exits and
their entrances;
And one man in his time
plays many parts,
His acts being seven ages.

—William Shakespeare

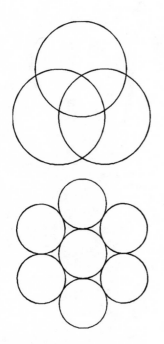

We encountered seven sections formed by Borromean Rings, and as six circles around one, but not as seven-sided figures in nature.

Only Seven Numbers exist. The ancient philosophers did not consider one, two, or ten to be actual numbers but rather their source and result, so the Dekad contains only seven numbers: three, four, five, six, seven, eight, and nine.

6×7 equals $7 \times 8 \times 9 \times 10$ (equals 5040). As a chasm, with seven absent, $1 \times 2 \times 3 \times 4 \times 5 \times 6$ equals $8 \times 9 \times 10$ (equals 720). Whether the value of seven is present or absent, its location serves as a pivot balancing the ten. No other number or position within the Dekad does this.

The ancients referred to seven as the "virgin" number. It is untouched by other numbers in the sense that no number less than seven divides or enters into it, as two divides four, six, eight, and ten, three divides six and nine, four divides eight, and five divides ten. Seven was also considered childless since it produces no other number (by multiplication) within the ten, as two produces four, six, and so forth.

To the ancient philosophers, myths were like our scientific and mathematical formulas. The gods and goddesses represented *principles* with certain properties and attributes associating them with mathematical archetypes. The relationships between gods and goddesses corresponded to the relationships among numbers and shapes, arithmetic and geometry.

The mythology surrounding virgin goddesses (Neith, Athena, Minerva) often elucidates the principles of the virgin Heptad.

The myth surrounding the "birth" of the Greek goddess Athena, or Pallas Athena Parthenos, was explained with the construction of a heptagon. Athena presided over war and wisdom, settled disputes, upheld cosmic law; she was first to teach the science of numbers, the arts, and trades; she invented cooking, weaving, spinning, the plough, ship, flute, pottery, and chariot; and she was patroness of the city of Athens. But she was not born in the ordinary way through a womb. When Zeus was told that one of his children would overthrow him, as he had usurped the throne of his father Kronus, he swallowed his pregnant wife, Metis, goddess of measure, mind, and wisdom, whole. He soon developed a fierce headache. When the blacksmith deity Hephaestus split his crown with an ax, Athena, mature and fully clothed for battle, with a mighty shout sprang forth from the cleft in his head. She remained virgin through life and, of course, bore no children. The origins of her name are unclear, but "A-thene" is thought to signify "I have come from myself" or *a-thanos*, signifying "deathless," "eternal." The birth of Athena is depicted on the east pediment of her temple in Athens, the Parthenon, over the entrance itself.

Mythology, religion, science, mathematics, and art were once part of an integrated system of philosophy. We can gain insight into this system by looking at the names of the gods and goddesses in their original language. In the ancient traditions of Greek symbolic mathematics the letters of the alphabet represent numbers. When the values for the letters of Athena's Greek name are added together they equal seventy-seven, a clear reference to number seven. The numerical value of the letters of her epithet, Pallas ("maiden"), combine to equal 343 (equals $7 \times 7 \times 7$), the volume of a cube whose sides are seven units long.

Statue of Athena from the Acropolis displays on her chest a gorgon head enclosed within a heptagon, associating sevenness with the virgin.

The letters of Athena's title Parthenos ("virgin") add up to 515. Five hundred fifteen is not a multiple of seven. But 51.4128571 . . . (nearly 51.5) degrees measures the vertex angle at the corner of a regular heptagon, the seven-sided polygon. Use a calculator to divide the 360 degrees by each of the values one through ten. While every one of the ten, *except seven*, divides 360 without remainder, only the seven-sided polygon presents an endless decimal and an unmeasurable, elusive angle from its center to its corners. Assiduously maintaining virginity, the Heptad cannot be captured, that is, made manifest, by arithmetic or geometry, the mathematical creating process. It appears to the geometer as an illusion, seeming to unfold but not uniting with any suitor among the many divisors of 360.

It's well known that the regular heptagon is the smallest polygon that cannot be constructed using only the three tools of the geometer, the compass, straightedge and pencil, the tools that mirror the methods of the cosmic creating process. In other words, an exact heptagon is not (and cannot be) "born" like the other shapes through the "womb" of the *vesica piscis*. This explains why the virgin seven cannot be entered (divided), cannot produce numbers within the Dekad, and cannot be captured: simply because it is not born.

Use a calculator to divide each number one through ten by seven. They each yield the same result: the sequence of digits 1–4–2–8–5–7 cycling endlessly, although they each

Among all the shapes, only the heptagon cannot be captured precisely.

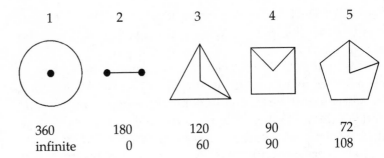

	1	2	3	4	5
Center angle	360	180	120	90	72
Vertex angle	infinite	0	60	90	108

begin with a different digit. Six digits, like the six days of the week, are set in endless motion around the unseen Sabbath.

The common saying "at sixes and sevens with each other" refers to seven's aloofness, its virginal unwillingness to mix with other numbers.

A carefully constructed mathematical canon underlies mythological relationships. The archetypal principles of arithmetic and geometry corresponded to deities in a system more widespread than is presently suspected, one that wove together the arts, sciences, and religion of ancient civilizations.

In the ancient "mythmatical" canon, each god and goddess with his or her titles and attributes represents a length (77), an area (spread by an angle of ~51.5 degrees), and a unit volume (343). The appropriate units were applied to designing two- and three-dimensional sacred objects, paintings, reliefs, statues, monuments, and temples dedicated to that deity. Temples to Athena, Minerva, and Neith were designed and constructed around the number seven. This tradition continued in Renaissance music, which called for seven voices when singing of the Virgin Mary.

Seven doesn't form finite relationships with other numbers. When any number is divided by seven it always leaves an endlessly looping trail of repeating decimals, 142857, mysteriously omitting 3, 6, and 9.

$$1/7 = 0.142857\ldots$$
$$2/7 = 0.285714\ldots$$
$$3/7 = 0.428571\ldots$$
$$4/7 = 0.571428\ldots$$
$$5/7 = 0.714285\ldots$$
$$6/7 = 0.857142\ldots$$
$$7/7 = 1.000000\ldots$$
$$8/7 = 1.142857\ldots$$
$$9/7 = 1.285714\ldots$$
$$10/7 = 1.428571\ldots$$

CONSTRUCT APPROXIMATE HEPTAGONS AND HEPTAGRAMS

(1) Construct a *vesica piscis* and (2) a square in the upper half. (3) Draw diagonals to find the center of the square, and use

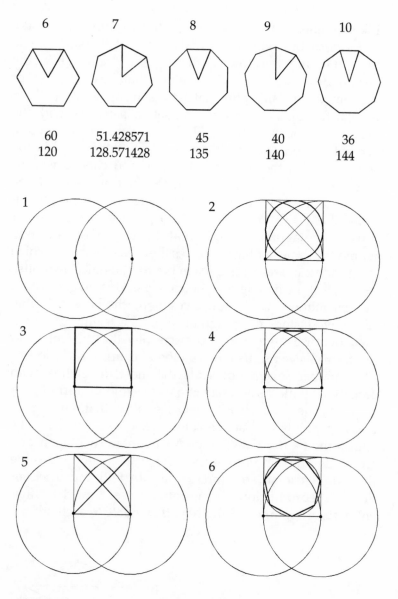

6	7	8	9	10
60	51.428571	45	40	36
120	128.571428	135	140	144

*A*ll *is uneven,*
And everything is left at
six and seven.

—Shakespeare

The Birth of the Heptagon.

it to inscribe (4) a circle within the square. These are symbols of heaven and earth, Monad and Tetrad. (5) Note the two points where the small circle crosses the circles beyond the *vesica piscis.* Open your compass to span these two points. This is the length of one side of the heptagram. (6) Walk the compass around the circle, leaving a mark at seven points. Connect them to configure an approximate heptagon (7).

Like Athene, it does not emerge directly from within the *vesica piscis* but pops out from the top of the small circle beyond it, from the crown of the circle, the Monad, from the head of Zeus, her "father." You'll find that the final "step" of the compass doesn't quite match the starting point. A "true" regular heptagon cannot be precisely constructed with the geometer's tools; it always eludes us. Seven is therefore a symbol of eternal rather than manifest "things." Walk the seven steps in the opposite direction and choose as your actual point a mark in between the first and second set. (8) Connect every other point to make one type of heptagram star. (9) Connect every third point to make another type of heptagram star. (10) Both heptagram stars drawn within the same heptagon creates the Web of Athena, which the goddess of weaving spins in her mythological exploits.

Gaze at the heptagram stars a while. Do you get a sense of their motion? Glide your eyes along the lines until you can move faster. A heptagram imparts a feeling of movement because our eye attempts to resolve the unevenness of the form. We sense its elusive eternal character as motion. The number seven represents the most direct link with archetypal patterns accessible to us in symbolic form.

Construct the Web of Athena. Doodle with it. Connect its crossing points and shade some areas. Get a feel for the angles and patterns of sevenness. Cut out the heptagon, and crease its lines on both sides. Fold it to create interesting three-dimensional structures. I like to fold constructions done on reproductions of ancient art to see how the different parts are composed and how they relate to each other.

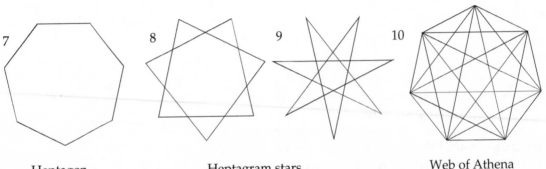

7   8   9   10

Heptagon   Heptagram stars   Web of Athena

Unlike Mother goddesses, such as Isis and Demeter, who are represented by the "solid" earthy square, the heptagonal virgin goddesses do not nourish us because they are not present in quite the same way. We glimpse only a specter of the virgin Heptad through our approximations with number and shape; she cannot be reached by worldly means.

MAKE THE HEPTAGON'S RELATIONSHIPS AUDIBLE

The Web of Athena is composed of only three lengths: the side of the heptagon and two types of length between points. We can make their abstract mathematical relationships more comprehensible by making them audible. Construct a large heptagon on a piece of wood, and hammer three nails halfway into the wood at three points as shown. Tie and tighten a piece of guitar string or nylon fishing line around them. Pluck them and listen. Find their tone sequence from high to low. Listen to the intervals they make, the jumps between them. Are the jumps the same or different? Pluck each two strings simultaneously, and listen for the rhythmic beat arising in their differences.

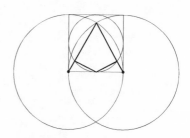

Another method: Construct a triangle and square within the *vesica piscis* to weave the principles of Monad, Triad, and Tetrad.

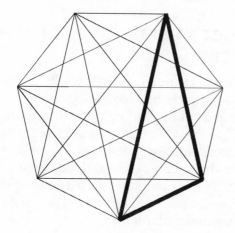

A 7-sided figure is
impossible to draw
With perfect mathematical
precision;
And if you try to do it you
are absolutely sure
To find your efforts treated
with derision.
And yet there are
philosophers who readily
declare
That nothing in this world
is really true,
And so I've drawn a
heptagon by triangle and
square.
For any human purpose it
will do.

—John Michell

APPROXIMATE A HEPTAGON WITH THIRTEEN-KNOTTED STRING

When the ancients wanted to construct a heptagon as the foundation of a temple they began with a loop of rope having thirteen equally spaced knots. To replicate this construction on paper tie thirteen equally spaced knots in a loop of cord or string. Form a triangle with sides of four, four, and five by stretching the loop around three thumbtacks. Use your straightedge to draw a straight line from the fourth knot on the longest side to its opposite corner. This makes one side of an approximate heptagon. Remove the two upper tacks, and rotate the triangle around the remaining central tack so that it continues where the previous triangle left off. Again mark the fourth point on the long side. Continue to rotate the triangular loop and mark each fourth point on the long side until the seven sides of the heptagon are created.

TIE A HEPTAGONAL KNOT

A seven-sided knot can be tied in one long, thin strip of paper. Follow the path shown in the illustration. Carefully pull the strip tight, pressing it flat, keeping the sides as equal as possible. Its underlying design reminds us of Borromean Rings, a trefoil knot, and an open lotus. Hold the completed knot up to the light to see the Web of Athena within.

CONSTRUCT A HEPTAGON WITH SEVEN TOOTHPICKS

By trial and error, lay out seven toothpicks as shown. Be sure that the two toothpicks making the top corner angle aim precisely at the ends of the bottom horizontal toothpick. Fourteen (= 2 x 7) of these triangles rotated around a point make a beautiful, more intricate variation of Athena's web within it.

Learn to draw heptagons and heptagrams freehand. Doodle with them to become attuned to the principles the Heptad holds. Connect points and extend lines to disclose further heptagonal patterns.

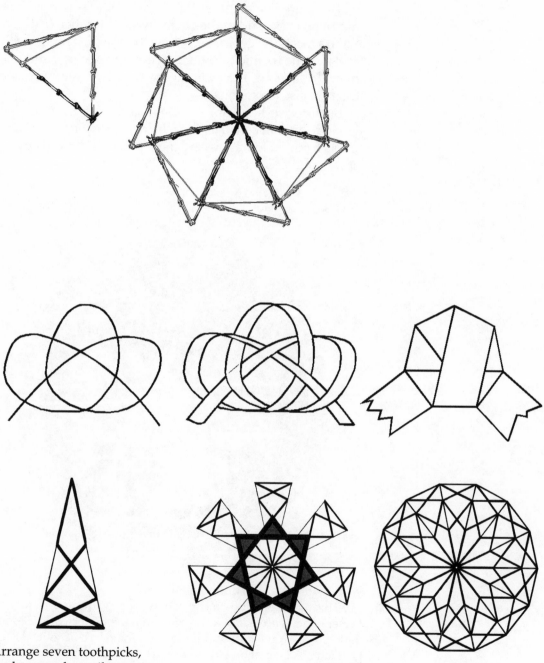

Arrange seven toothpicks, unsharpened pencils, straws, or nails to form this section of a heptagon.

These are just some of the many methods people have employed in pursuit of the elusive virgin number. Just as no method using the geometer's three tools, which replicate the "tools" of the cosmic creating process, light, energy and matter, will make an exact heptagon, neither can nature construct precise heptagons.

But precise heptagons are not necessary in natural or symbolic human constructions. In fact, they can *only* be seen "imperfectly." Heptagons are most powerful when they are represented this way.

Weave Athena's Web upon Classic Art

This painting within a Greek bowl depicts the virgin Kore, or Persephone, with Hades kidnapped into the underworld. Seven dark crosses appear around the rim, giving us a clue to its design geometry. A Heptagon has been constructed around the circle with its corners halfway in between the seven crosses. Place tracing paper over the picture, and connect the heptagon's corners to produce the regulating lines that compose the scene.

ENCHANTING LYRE

The math and myth of seven, the Heptad, are intimately related to those of twelve, the Dodekad. Both have in common the interplay of Triad and Tetrad, triangle and square. That is, $3 + 4 = 7$ while $3 \times 4 = 12$. These simple equations may not seem like much, but they are responsible for myriad cosmic structures and relationships. We saw how the twelve-part geometry of the squared circle (chapter six) with twelve moons around a thirteenth at the center holds the canon for the structure-function-order of a settled society. Within it are heptagram stars representing a different kind of structure-function-order, that of the wandering, nomadic life. While twelve has relations with all numbers within the Decad, seven has none. Seven is the hidden, unborn, eternally elusive side of twelve.

Seven and twelve are associated so often in myth and legend that it becomes obvious that they refer to a system of symbolic philosophy. It was noted earlier that twelve vultures showed Romulus seven hills on which to found Rome. The Greek pantheon of twelve deities was represented on earth by the seven peripatetic wise men. Most widely noticed was the association between the seven nomadic *planets* (from the Greek for "wanderers") whose courses change periodically against the backdrop of twelve fixed zodiacal constellations. In the Kabalah, the twenty-two letters of the Hebrew alphabet are divided into three groups, of three, seven, and twelve, representing, on one level, our deeper self, the solar system, and the galaxy.

Seven-twelve also describes the essential relationship between the modern seven-note diatonic and twelve-note chromatic musical scales. The chromatic twelve-scale *includes* the diatonic seven-scale plus five sharps or flats between them, giving the scale shadings of musical color ("chroma").

C C# D D# E F F# G G# A A# B

The modern piano of eighty-eight keys displays seven and one-third chromatic scales within the approximately ten octaves of human hearing.

The Chinese, Polynesian, and Scottish five-note penta-

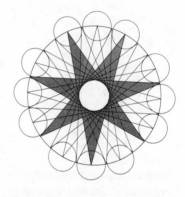

Geometry of the twelve-part celestial canon, as the piano keyboard, relates to sevenness since $3 \times 4 = 12$ and $3 + 4 = 7$.

tonic musical scale, which most easily produces the feel of yearning, is made of five of these twelve notes.

The people of all higher cultures have felt deeply that the world is ordered. In ancient times the design of a musical scale was part of a sacred philosophy reflecting the world's inherent order, its tones not chosen accidentally or purely by ear. The musical scale was another way to represent a model of the universe and served as a tool for enhancing peoples' harmonious relation with it.

Western civilization's seven-note *diatonic* (from the Greek "across the tones") musical scale (the piano's white keys) has been used from time immemorial. In ancient times it was traditional to arrange the strings to play the scale downward, as if it were descending from heaven. The modern names of the seven familiar notes in descending order, DO–SI–LA–SOL–FA–MI–RE–DO, were proposed by Guido d'Arezzo, inventor of the musical staff, around 1000 A.D. These popular names are only the first letters of Latin words whose translation reveals a cosmological structure derived from an earlier age:

DOminus	"Lord"	Absolute
SIder	"Stars"	All Galaxies
LActea	"Milk"	Milky Way Galaxy
SOL	"Sun"	Sun
FAta	"Fate"	Planets
MIcrocosmos	"Small universe"	Earth
REgina Coeli	"Queen of the Heavens"	Moon
DOminus	"Lord"	Absolute

The seven-note scale is meant to model the hidden side of the macrocosmic design, the universe ruled by mathematical harmonies of music. The scale structure implies that the universe emerges from absolute divinity, descends through a seven-stage celestial hierarchy, and returns to absolute divinity. In order to create music the string is bowed back and forth by the motions of the generative Dyad between the extremes of harmony and discord, love and strife, light and shadow.

The ancients designed and used musical scales to play the harmonies of the heavens, the music of the spheres

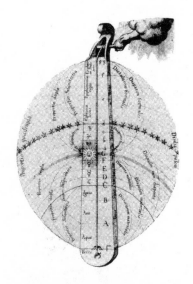

Whether the cosmos is represented as a musical scale as in ancient times or as the modern sonic and electromagnetic spectrum, they both depict a universe based on vibration.

See deep enough, and you see musically; the heart of Nature being everywhere music, if you can only reach it.

—Thomas Carlyle

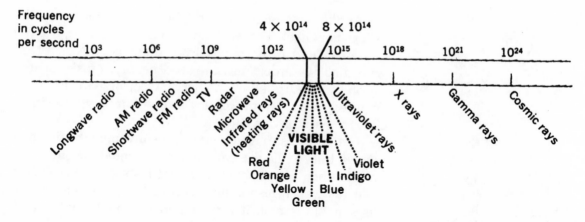

Electromagnetic spectrum.

pleasing to both gods and humans. Music was meant to allow the higher principles to enter our lives through our sense of hearing and our emotion.

Musical strings were considered imperfect representations of archetypal ideals and pure mathematical motions. When vibrating, these strings were thought to resonate with the archetypal tones, manifesting cosmic principles. Thus, music was seen as having great power for producing harmony on earth.

Music of the lyre was used to accompany and enhance the voice. But it was part of a larger administrative and therapeutic system involving colors, fragrances, flavors, textures, herbs and plants, minerals, metals, myths, agriculture, meteorology, and astronomy.

It was well known throughout ancient civilizations that music bypasses the intellect with the power to manipulate passions and emotions. In this way music was used to heal. As we find by being very quiet within and observing ourselves, sound has an effect on our inner substance. If we were more aware of the effects sounds have upon us we would be more careful about the frequencies we expose ourselves to.

Healing or therapeutic music was attributed to Egyptian, Babylonian, Indian, Chinese, African, Native American, and Persian priests, to Pythagoras, and the mythical Apollo and Orpheus.

Apollo

Apollo ("moving together"), brother of Athena and the only major Greek deity with no Roman counterpart, was the Olympian god of harmony. Everything he was involved with displays order, measure, and beauty: music, poetry, prophecy, medicine, art, archery. According to myth, he acquired the lyre from Hermes and with it taught the principles of harmony on earth.

The Three Fates were said to have invented the first five vowels, but Apollo was said to have invented the other two vowels (long o and short e) so that his sacred lyre would have a vowel corresponding with each of its seven strings. His son Linus took it further and invented rhythm and melody.

Apollo presented his lyre to Orpheus, who, myth tells us, used it to enchant people, plants, animals, and the weather. By its tones Orpheus guided his wife, Eurydice, from the darkness of the underworld into light. He is said to have made trees and rocks dance to his music throughout Greece, where they remain to this day in the merry patterns he left them.

Pythagoras is said to have calmed a raging bear and soothed the murderous passions of a drunken man by playing appropriate tunes on the lyre.

In ancient times the design of a musical scale involved the science of "mediation," the binding of two "opposite" tones by a third tone placed harmoniously between them. *Harmony* (from the Greek *harmonia,* signifying "fitting together") in music, as elsewhere in the cosmos, comes from the reconciliation of opposites by a third element, bringing them all a new unity. The ancients saw that music theory holds the principles of social harmony *and* discord.

Numbers themselves arise from opposites, from the interplay of Monad and Dyad. Geometric shapes emerge from opposing circles of the *vesica piscis.* A tone is created when a motionless string is plucked by a moving finger. In Greek myth the goddess Harmonia was born from opposites, her mother Aphrodite, goddess of love, her father Ares, god of war. Harmony is their balance. Through music the ancient philosophers could hear the relationships among the gods and goddesses.

Sound is simply vibration, but the agreeable and disagreeable sensations produced by music are due to tones being combined harmoniously or discordantly. Cosmic harmonies played with earthly strings bring the archetypal principles to earth, thus making music therapeutic.

The most famous musicosmic scale, the one Western civilization uses today, is derived from the scale developed and taught by Pythagoras before 500 B.C. Undoubtedly he learned the mysteries of music in Egypt, Chaldea, and India, but he was the first to teach its principles outside the temples. Pythagoreans called the harmonious cosmic relationships the "music of the spheres" and taught that music is based on the mathematical relationships, found in nature, and can be expressed by ratios of simple whole numbers.

If you have a guitar or other stringed instrument, or even a rubber band stretched between points, you can try these experiments to replicate the cosmic harmonies and forces in the safety of your own home.

First, pluck a string of any length, thickness, and tension. The string is a Monad. All subsequent harmonies begin and end in this unity. Listen to the tone without naming it. Observe the shape of the vibrating string. It resembles a *vesica piscis*, a womb whence sound is born. But is sound really *born*? Sound is a transitory wave of air and doesn't stay long, just as the elusive heptagon isn't really constructed.

Every structure in the universe is attuned to a level of the cosmic scale. Tap any object and listen to the tone it makes. It's telling you its natural name. Each configuration has its own note, its vibration by which it sings to the cosmos.

Now estimate by eye (don't count frets) where you think the exact middle of the string is. Press (or put the bridge) there. Pluck each "half" length and listen. You'll know if you've found the exact middle because if you have, each half will sound the same. Now find the center by *listening*, moving your finger until both halves sound identical. Most people can find the middle of a string much more accurately by ear than by eye.

Pluck the entire string again, and you'll recognize it as the same note as each of the halves, but lower in pitch.

On this account Pythagoras kept a lyre with him to make music before going to sleep and upon waking, in order always to imbue his soul with its divine quality.

—Censorinus (Third century A.D., Roman scholar)

Pythagoras performing the first experiment in physics, the mathematical tuning of the musical scale. The weights of the strings result in different tensions and tones. The numbers representing weights can just as well be the lengths of string with identical tension.

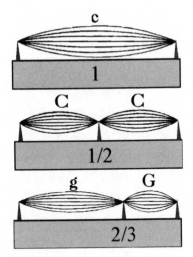

Vibrating string lengths represent the fundamental tone (C), its octave (c) , and the harmonious fifth (G), the "chord of triumph," between them.

*M*usic is a secret arithmetical exercise and the person who indulges in it does not realize that he is manipulating numbers.

—G. W. von Leibnitz

These opposites of full string and half string define the bounds of the cosmos. Monad and Dyad are the parents of all tones just as one and two are the parents of number, and the circle and *vesica piscis* give birth to all geometry. Although the sounds produced by whole and half-string sound different, they are alike, being one octave apart. But unless they form a relationship *in between* them they produce nothing new, always the same tone, the One.

Harmony is the child born between these opposites. She appears on the string at the point that divides it into two *unequal* parts so that *plucking each part produces the same note as the other part*, but different in pitch, an octave away. Find this dividing point by moving your finger (or bridge) along the string, plucking each part, and *listening for the same note arising in each*, one high and one low. This is the tone of concord between the Monad and Dyad that the ancients sought as the foundation on which to build the musical scale. Internal self-agreement is the essence of harmony in music, nature, and human affairs.

Pythagoras found this "balancing" point at two-thirds the full length of the string. A division here creates two parts, one of which is twice (two-thirds) the length of the other (one-third). The two parts sound the same as each other, also an octave apart. They replicate between themselves the doubling relationship of their parents, the Monad and Dyad. This division into "thirds" introduces the Triad, a birth, a child of the essential unity different from its parents yet in harmony with them.

Pluck the full string, then two-thirds of the string, and listen. No matter what the original tone, its two-thirds tone has a certain feel to it, in relation to the whole string. It is traditionally known as the "chord of triumph" for the inspiring way it moves everyone who hears it. We respond not to the tones themselves but to their *relationship*, their musical *interval*.

By continually subdividing each two-thirds section by another two-thirds we create a sequence of shorter strings and higher tones, each building upon and maintaining the same harmonious relationship with the notes before and after it, always the chord of triumph. If you are careful you will find, as Pythagoras did, that after the *seventh* division

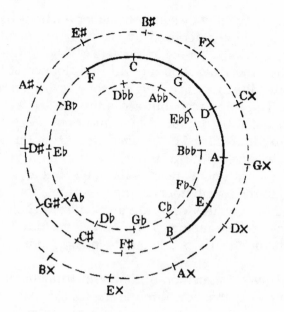

Pythagorean tuning. Endless scales can develop as "fifths" (clockwise) by continually dividing a string at two-thirds (or extending by four-thirds); or as "fourths" (counterclockwise) by continually dividing at three-quarters (or extending by three-halves).

the notes repeat the earlier cycle, but sharper. He also found that *after the twelfth* division the original note was virtually sounded again, although slightly flatter. These twelve tones spread across seven octaves complete the cosmic picture, although the sequence of notes is not alphabetical. These are the twelve notes of the piano's chromatic scale represented by its seven white and five black keys. They lead us to the enchanting virgin. Her song is the mathematics and physics of music itself.

This process of continually dividing a string by two-thirds to produce new higher tones is called the "circle of fifths," although it is more like a "spiral of fifths." "Fifths" refers to the fact that each subsequent division at two-thirds produces the *fifth note ahead* in the musical sequence A–B–C–D–E–F–G. In other words, if a length of string sounds note A when plucked, then two-thirds will produce a higher note E; two-thirds of that will sound a higher B, and onward, always five letters ahead. The fraction two-thirds was so important to the ancient Egyptians, who successfully integrated music within their civilization, that they honored it as the only fraction to have its own glyph. All other frac-

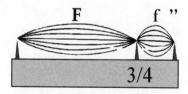

The musical "fourth." Dividing a string at three-quarters of its total length produces the "fourth" note ahead in the scale. Continually subdividing into three-quarters produces the same twelve-tone scale as the spiral of "fifths" but less rapidly and in the opposite direction.

tions were denoted by sums of smaller fractions having a 1 as the numerator depicted by the "mouth" glyph, like a vibrating string, over the numeral (e.g., $3/4 = 1/2 + 1/4$).

When the same "spiral of fifths" is looked at from the opposite direction, each higher tone is the *fourth* letter ahead. The same scale but ascending, called a "spiral of fourths," can be produced by continuously *extending* the string by four-thirds, or by *dividing* it at its three-fourths length. A tone's "fourth" is the inverse of its "fifth." On the spiral, the fifth note ahead is the same, but an octave different, of the fourth note behind. A string's fourth and its fifth added together comprise an octave, a whole comprehending both harmonies. The tone, octave, fourth, and fifth hold the mathematics that builds the harmonious musical scale.

This was the method of tuning instruments used from the time of Pythagoras through the Middle Ages and into the Renaissance. It produces more *concords* among its constituents than any other selection of tones. Musical instruments tuned to the ancient Pythagorean scale sound brighter, as an old fresco looks when cleaned of centuries of grime. Some native peoples around the world and even some modern rock groups tune their instruments the ancient way.

These twelve notes were the ones held by each choir of the twelve provinces of classic settled societies. The spiral of fifths represents the zodiacal year in twelve phases, twelve processes of life, twelve spiritual attributes.

Musicians eventually noticed that the thirteenth step *should* bring us back precisely to the starting note seven octaves later, but in reality it is slightly flat. The actual vibration is only 98.65603. . . percent of the archetypal ideal, a tiny difference termed the "comma of Pythagoras." The ancients knew this tuning to be very powerful, literally enchanting and therapeutic. But when orchestras developed in the seventeenth century, musicians wanted to tune their instruments *precisely* with each other. The problem with the ancient tuning is its difficulty modulating between keys. So the comma of Pythagoras was distributed evenly among the twelve notes, flattening each by one-forty-eighth. This system, which tunes the modern piano, is called the "equal tempered scale." While the ancient names of the notes are the same as those found on the piano, the tones themselves are actually different since the piano's strings have been "equally tempered." The result is that the piano's thirteenth tone will precisely match the first tone, an octave higher. While the slight flattening of each note may not seem like much, it saps the power and therapeutic value in the tiny differences, the "lawful inexactitudes," from the archetypal ideal that cannot be "captured." The equal tempered scale is an attempt to capture the eternally elusive Heptad.

The following geometric construction will guide you to the ancient musical scale that tuned the therapeutic lyre of Apollo, Orpheus, and Pythagoras.

THE GEOMETRY OF MUSIC

Reproduce the Musical Scale of the Lyre of Apollo, Orpheus, and Pythagoras

In ancient days a scale on the lyre was played by stroking upward, starting with the shortest string, the highest note, at the bottom of the lyre. The earthly music was seen as a mirror image of the heavenly ideal descending from above. The therapeutic lyre of Apollo, Orpheus, and Pythagoras placed high E at the bottom. As one stroked the strings upward, the tones "descended" to earth through E–D–C–B–A–G–F. An eighth tone, a lower E, was added to complete one "octave" (see chapter eight).

How do we begin to construct this scale? We look to "mythmatics." In the alphanumeric system of gematria, the name Apollo has a value of 1061. The name Zeus, the central Olympian deity, has a value of 612. The name Hermes, the messenger who invented the lyre, has a value of 353. The relationship of Apollo to Zeus can be expressed mathematically as $1060/612 = 1.732$, the value of the square root of three ($1.73205 \ldots$). And the relationship between Zeus and Hermes is virtually the same ($612/353 = 1.736$). Root three also describes the relationship between the two axes of the *vesica piscis*, which is the proportion within the triangle.

*W*hat passion cannot
Music raise and quell?

—John Dryden (1631–1700,
English poet)

Ancient myths were represented by geometric constructions. Apollo/Zeus = 1060/612 = 1.732 = the square root of three, the relationship of the axes of the *vesica piscis*, the crossing opposites through which the musical scale is born.

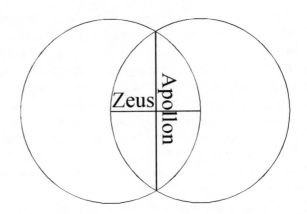

Thus, Apollo's harmony, like the triangle, is born from the interplay of Monad and Dyad, same and opposite.

With this hint all we need are the geometer's three tools and a *vesica piscis*. No numbers, ratios, fractions, decimals, or arithmetic are necessary. The lyre's concords are found purely through the harmonic geometry of a turning compass.

If you actually want to build a device to make the cosmic harmonies audible, you can wrap a guitar string, long piano wire, or nylon fishing line tightly around properly placed nails or screws in wood or on a sounding box. These lengths can also be used to determine the hole placements of a flute, tube lengths for panpipes and wind chimes, xylophone tones, or identical glasses filled with water to different heights.

Follow these steps, a metaphor for the stages through which the cosmic creating process unfolds its seven-part harmonies.

1. Construct a circle whose diameter represents a length of string which sounds the note B when plucked. Beginning at tone B conveniently puts off the sharp notes until the seven notes of the scale are produced.

2. With the compass open to the same radius and centered at point B, turn another circle and create a

vesica piscis. Extend the diameter to meet the circumference of the new circle at the length of string which sounds the lower note E, a musical "fourth" from B.

3. Draw a vertical line between the points of the *vesica piscis.* Center your compass where the line crosses the diameter and open it to point E. Turn a circle to enclose both the smaller circles.

4. With the compass open to the same radius and centered at point E, turn a circle and create a larger *vesica piscis.* Extend the diameter to determine the length of string whose tone sounds the lower note A.

5. Repeat steps (3) and (4) four more times creating larger and larger *vesicas pisces* which give the lengths for lower notes D, G, C, and F. If you wish to construct the longer lengths for the full twelve-note chromatic scale, continue the process to create notes B♭, E♭, A♭, D♭, and G♭.

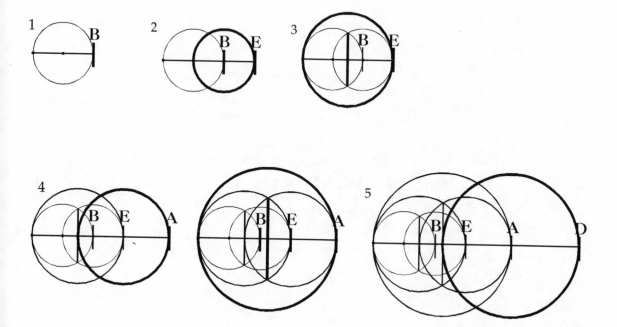

○

*F*or the true knowledge of
music is nothing other
than this: to know the
ordering of all separate
things and how the Divine
Reason has distributed
them; for this ordering of
all separate things into
one, achieved by skilful
reason, makes the sweetest
and truest harmony with
the Divine Song.

—Hermes Trismegistus

Each larger circle extends the previous diameter by one-third; that is, when the scale is descending, each string is four-thirds its original length. Musicians call this interval a "fourth" since each next note is four letters ahead alphabetically. When built in the other direction as an ascending scale, each note is a "fifth" ahead. In either case the result is the same spiral.

Gaze at the completed geometry until it seems to become a tunnel of expanding *vesicas pisces*.

When these lengths of string are plucked sequentially or simultaneously, each of the twelve tones is harmonious with those directly before and after it.

If the geometry on this page were audible and we could hear these tones as they are constructed, we would hear them growing rapidly lower. These twelve tones span seven octaves. We need to gather them into a single octave within the reach of the human voice. This, too, is done by geometry, lowering notes by doubling short lengths. Every doubling of a string's length lowers the octave but maintains the note. We will use this technique to bring all the notes into one octave,

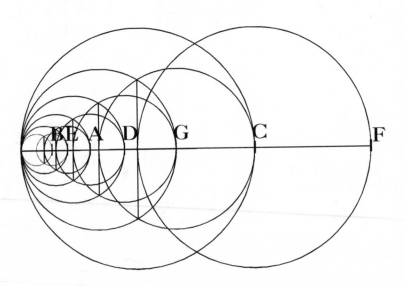

These seven notes span three octaves.

in an alphabetic sequence within the last circle we drew.

The ancient Greeks recognized seven modes of musical scale, each a sequence of twelve notes along the spiral of fifths (or fourths) but each beginning with a different tone. The modes were said to resemble seven human "moods" and would be played for different therapeutic and social purposes. The Dorian mode was considered masculine, the Ionian feminine. Each mode also had a different architectural representation known by the same name.

The Pythagorean sequence, the Dorian scale, begins with note E and descends: E–D–C–B–A–G–F (E'). To refashion our construction into this sequence we must double different strings' lengths once, some twice and some even three times. To double any length, just open the compass from the end point where all circles meet to its mark and walk this length along the line. The placement of notes F and C need not change. Double the lengths for notes G and D. Do two doublings, or walk four lengths, to lower the notes A and E. And double, or walk, the shortest distance representing note B eight times. This will finally produce the ancient sequence of string lengths for a Pythagorean scale, a vibrating replica of the relationships within the expanding sevenfold cosmic structure.

The word "music" comes from the Greek *mousike* ("muselike"), which derives from the Egyptian glyph *mes* ("birth, generation").

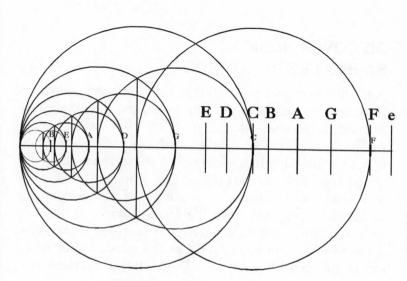

There is geometry in the humming of the strings.

—attributed to Pythagoras

Doubling the lengths of shorter strings retains the note but lowers its octave, bringing all notes within the same circle and octave.

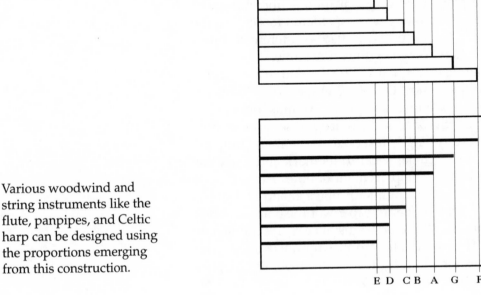

Various woodwind and
string instruments like the
flute, panpipes, and Celtic
harp can be designed using
the proportions emerging
from this construction.

DISCOVER MUSICAL HARMONIES IN PAINTING

Botticelli's painting of Venus depicts the Roman goddess of
love, beauty, and affairs of the heart, the Greek goddess
Aphrodite, mother of the goddess Harmonia. Venus/
Aphrodite was not born in the usual way but is shown here
in her mythical habitat, arising from the foam of the sea.

Botticelli, like many other artists, sculptors, and architects over thousands of years, composed his works using
harmonious proportions of the musical scale. In this way
their paintings, reliefs, statues, and temples were music
made visible, as pleasing to the eye as music can be to the
ear. In the case of temples and cathedrals, specific sacred

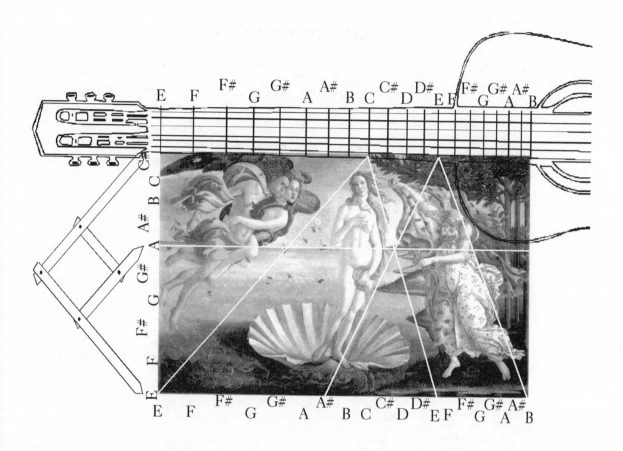

hymns were sung in keys associated with that deity or saint resonant with the architecture.

This painting, a golden rectangle (see chapter five), is divided along its perimeter into the tones of a full musical octave (E to E) and a "fourth" beyond it (B). Botticelli particularly emphasizes the concords of the "fourth" (G and E) and "fifth" (A). To discover for yourself how they guide the placement of figures in the scene, place a piece of tracing paper over the page and connect the rim's emphasized points with straight lines. Note how a line from the C above to each bottom corner divides the canvas into three triangles containing the winds, Venus, and nymph. Connect the significant points in other ways to explore the painting on your own.

SEVEN CRYSTAL SYSTEMS

If you wish to employ the Heptad's principles when constructing a universe, you should not expect to create any seven-sided objects. For example, among the myriad crystals, from quartz to calcite, lead to gold, none are seven-*sided*. The reasons are simple. The heptagon's unborn angles cannot completely cover a flat surface or fill three-dimensional space without leaving gaps, so nature has no use for it as structure.

Instead, the Heptad pokes into the world as a seven-step process, seven independent stages or aspects of a whole.

There may be no seven-sided crystals, but the Heptad appears as seven major groups or "systems" into which all of nature's beautiful jewels belong. Their differences are due to seven different possible relationships among their axes and angles.

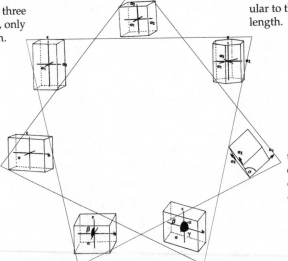

Cubic crystals have three mutually perpendicular axes of equal length.

Hexagonal crystals have four axes, three of which are of equal length and lie on a plane with an angle of 120 degrees between them. The fourth axis is perpendicular to the three and can be of any length.

Tetragonal crystals have three mutually perpendicular axes, only two of which are equal length.

Orthorhombic crystals have three mutually perpendicular axes, each of different length.

Rhombohedral crystals have three axes of equal length but an angle of other than ninety degrees between them.

Monoclinic crystals have three axes of different length, only two of which are perpendicular.

Triclinic crystals have three axes of different length, none of which is perpendicular to any of the others.

Look at these pictures, beginning with the cube, universal symbol of "earth" and the "solid" phase of matter. The cube is the basic volume from which the other crystals extend. We usually tend to notice a crystal's smooth faces. But look at the cube from the center outward. See its three mutually perpendicular axes of equal length. Look at them until you can feel how space extends along each axis to manifest the volume. Go counterclockwise around the seven and view each this same way, getting a feel for how the changing relationships among the axes, their lengths and angles, determine the seven different possible systems. In each system one relationship changes, forming a progression of proportions among the lengths and angles of their axes. In the cube three axes are the same length and perpendicular. In the tetragonal system only two are the same. Next, none are the same length, but they are still perpendicular. Then the angles change, until seven possible "systems" occur, seven close-packing possibilities of atomic lacework.

Seven stages, or distinct steps, can be seen elsewhere in nature, especially where a doubling takes place. We saw this in the musical scale's seven tones occurring between the full length of string and its half. An example as close as your body is the process by which it grows, the division of one cell into two. This process of mitosis (see chapter two) is simply a transformation along seven stages or intervals. When the eighth step is taken, the octave is achieved and the one cell doubles to become two.

Gambling, like warfare, has always been an impetus for the development of mathematical ideas. Mathematicians and computer scientists studying the possibilities within a deck of cards have found that when the pack is shuffled it is simply being split into two parts, which are recombined through interleaving. They asked, "When is a deck of cards completely shuffled and random so that all cards have equal probability of being dealt?" Mathematicians have found that . . . you guessed it, *seven* shuffles is just right to create randomness. Fewer shuffles don't mix the cards enough, and more than seven shuffles is superfluous. Seven steps again complete a whole.

Seven-step rituals appear in every culture and religion. Reverence for this number, from the Seven Stations of the

Nature delights in the number seven.

—Philo Judaeus (c. 20 B.C.—c. 40 A.D., Alexandrian philosopher)

Cross to the seven basic ballet steps, has occurred independently around the world and across time. Natural expressions of the Heptad are not as obvious as the results of other archetypal numbers and shapes because they must be seen as a sequence of relationships. But they can be discerned if we can learn to see, hear, and sense differently, to recognize the processes we are part of, especially those sent to us in elusive packages.

LIGHT FANTASTIC

The most elusive expression of Heptad is a rainbow. No more enduring than vapor and light, the Greeks personified this passing hue as the messenger goddess Iris on a mission for Zeus leaving her varicolored trail across the sky. When she is not in our sight we carry with us the daughter of Thaumatos (Wonder) as our iris, the colored part of our eye, the organ that perceives the wonder of light and color.

Iris

Around the world, in religion, myth, and legend, light is considered the purest expression of divine principles accessible by our senses. The Sanskrit word for "angel" is *deva*, literally "shining being," which gave rise to both "divine" and "devil." Light has always been a symbol of greater understanding, a link between the eternal and the worldly. That's why the rainbow has been depicted in myth as a road or pathway, a bridge, an archer's bow, a divine necklace, a good snake, and a promise of divine beneficence.

The wonder of sunlight and starlight is that it sets the universal speed limit at 186,282 miles per second, faster than anything else, traversing trillions of miles in absolute silence, yet it can be stopped by a mere outstretched hand. There's no impact since the mass of a photon, a "particle" of light, is zero. In fact, a photon ceases to exist when it stops, since its energy is absorbed by whatever it hits. In a sense, light isn't even there.

Yet upon this loom of starlight all matter precipitates. The cosmic creating process, represented by the equation $E=mc^2$ and symbolized by the geometer's three tools, describes how light "thickens" into archetypal energy patterns that further condense into configured matter in the form of the energy pattern (see chapter four). But a rainbow never reaches that most dense stage of materialization. It remains perceptible yet ever intangible, briefly revealing itself only as a sevenfold pattern of energy. A rainbow shows us the underlying pattern, the frame of the structure before solid walls are put on.

The sevenfold pattern is not limited to sound, crystals, and light but is characteristic of all vibratory phenomena. All sense perception (sight, hearing, taste, smell, touch) transpires by means of our registry of vibration and as such abides by the harmonies inherent in the sevenfold mathematics we hear as music.

The retuning of the ancient musical scale into equally-tempered notes came about during the same period in which Isaac Newton performed his famous experiment with a glass prism in a darkened room with only a narrow slit for sunlight to enter. To Newton's delight a shaft of sunlight entered the prism and emerged as a rainbow or chromatic spectrum spread across a wide swath.

Sunlight is "whole" in that it is a mix of many vibrations overlapping. Some waves are very long and some very short. Emerging from the sun they all blend into white. Sunlight is a unity, representative of the Monad. The glass prism refracts the sunlight, bending and spreading out its mixed

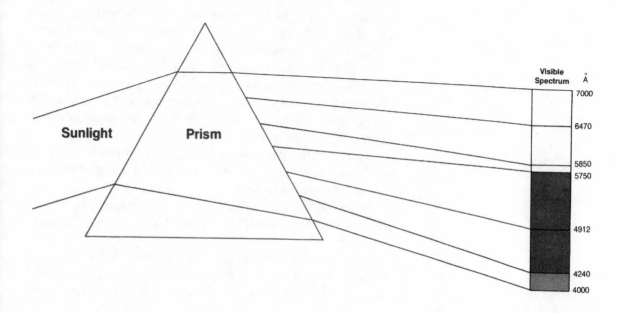

waves like a deck of cards spread across a table. Like Athena born from Zeus's head, the seven colors of the rainbow—red, orange, yellow, green, blue, indigo, violet—appear to spring forth "fully clothed." Red's waves are longest, violet's shortest.

Just as the intervals between the notes in the musical scale are not always the same, the color swaths of a rainbow are not of equal width but are proportioned like musical notes. Notice the different widths of color the next time you see a rainbow in the sky or around a garden sprinkler.

Many people have seen rainbows and some know the acronym for the color sequence "ROY G. BIV"; but few people know whether red or violet is on the top of the rainbow (don't take my word for it but check for yourself that it's red).

A rainbow will only appear to you if you're facing *away* from the source of light. The sun shines from behind and over you, does a *double* refraction within atmospheric water droplets, and the solar vibrations splay and wave into your eyes, where they are interpreted as color.

We claim to see rainbows and yet, like "constructed" heptagons, they're not really there; they don't manifest themselves "at" any location. People standing apart will see the same rainbow in different places. Even your two eyes see two separate rainbows. Verify this phenomenon by viewing a rainbow with only one eye at a time—it will seem to shift. Walk, and the rainbow will seem to follow your steps. Try as you will to find its end, it will elude you, eternal virgin that it is. The rainbow appears as an arc, but it's part of a full circle, which would continue underground if it could. The Greeks saw this as Iris' ability to travel between heavenly Olympus and the underworld.

The rainbow provides a transitory glimpse into eternal principles. Light touches everything. Anything can stop it, yet nothing can pollute it.

Since the rainbow seems to have infinite shades, why do we see exactly *seven* colors? As usual, the answer lies within ourselves.

Around 1800, the English musician-turned-astronomer William Herschel put a thermometer in the path of each color of a spectrum in a dark room. He guessed correctly

DESCARTES' DIAGRAM OF THE RAINBOW
(From his *Les Météores* of 1637.)

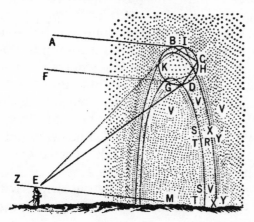

that they would have different temperatures. But to his surprise, the "empty" colorless space just *beyond* each end of the spectrum had a higher temperature than any of the colors. By thus discovering infrared and ultraviolet light he showed that the spectrum is a continuity of which we register only a narrow slice we label "visible light." But visible light is part of an endless range of vibrations extending to infinity at each end. Herschel discovered what moths, birds, bees, and other forms of life already knew: there are vibrations beyond these seven. Astronomers tell us that most of the universe (over 90 percent) is invisible to our eyes but can be detected with equipment sensitive to radio waves, microwaves, X rays, gamma rays, and different wavelengths for which we have other labels.

The clue to the rainbow puzzle is in the relative size of their waves. Each wave of light our brain interprets as "red" is 1/33,000 of an inch long. Violet's waves are about half that at 1/67,000 inch. Much as plucking a string and then another half its length produces musical notes identical in tone but different octaves, we see red and violet as the first and seventh "notes" of a "scale" just before reaching red's full doubling or "octave" wavelength past violet. Just as we can hear about ten octaves due to the spiral cochlea of our ear (see

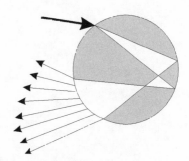

Sunlight refracts twice within spherical water droplets, changing direction and reversing the order of colors we see.

G	ultraviolet
F, F#	violet
E	indigo blue
D#	cyanogen blue
D	greenish blue
C#	green
C	yellow
A#	orange-red
G#, A	red
G	infrared

The physicist Hermann von Helmholtz devised a correspondence between the visible spectrum and musical scale.

chapter five), our neural network is prepared to register less than one octave of light, this particular seven-step slice of the endless electromagnetic spectrum. The range of colors is exactly seven because these are the only combinations possible with the tiny "rods and cones" on our retina. Rods perceive black and white but three kinds of cone, sensitive to red, green, and blue, combine their input into seven colors, like the seven areas within three Borromean Rings. Look at the screen of a color TV (while it's on)—with a magnifying glass if you have to—and you'll see a fine net of red, green, and blue holes responsible for the full seven-color spectrum we see, matching the receptors in our eyes.

In effect, colors are not "out there" but always within us. A spectrum is a colorless spread of vibrations of energy from which we make a picture.

Nature is consistent in her patterns. Oddly enough, the colors of the planets past earth happen to form the familiar spectral sequence: Mars–red, Jupiter–orange, Saturn–yellow, Uranus–green, and Neptune–blue. Perhaps the solar system is a refraction of light from within the galaxy and the planets show a spectrum arranged with the harmonies of a

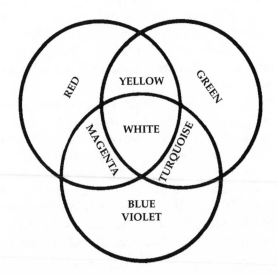

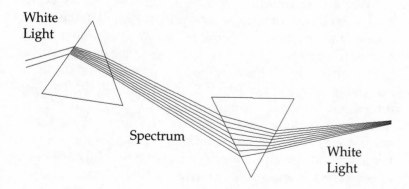

White
Light

Spectrum

White
Light

musical scale. The Pythagoreans were not far off when they spoke of the music of the spheres, a celestial choir harmonizing in octaves we cannot ordinarily hear.

There was another part of Newton's experiment, which is lesser known but which has profound implications. He took a second prism and directed the colored spectrum into it. To his amazement the colors reemerged from the other side of the prism reunited as a single shaft of whole white light. Just as the Monad creates all numbers yet never loses itself, just as the whole string retains its tone, so it is with light. Regardless of appearances, the source is never lost or far from us.

THE RAINBOW BRIDGE

In our day of proliferating self-help, self-awareness, and self-development workshops, books, tapes, videos, software, and multimedia technology, it may be difficult to remember that until fairly recently open discussion of the inner life has been taboo. For countless centuries the deeper understanding of our inner structure-function-order was kept within temple walls, mystery religions and oral traditions among small groups, which sought to avoid the wrath of religious intolerance and the misuse of their knowledge by others. The significance of the number seven to our inner life was kept disguised in mystical terminology, metaphor, and myth because it refers to a great secret—the knowledge of self-transformation.

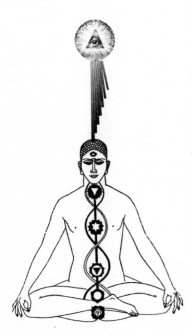

One, two, three, four,
five, six, seven;
All good children go to
heaven.

—Children's song

Worldwide tradition acknowledges humans as being composed in accord with natural principles and made in the image of the cosmos. A useful metaphor of our inner life is that of the prism and spectrum. Imagine the existence of a great, conscious, divine light beyond measure, the mysterious power of awareness itself, shining through the Triad of our Higher Self, as if through a triangular "prism." It refracts into seven "colors," seven centers of gravity, the ancient seven-tone "scale" of our soul. And upon this inner spectrum our body precipitates, ruled by the powerful hormones of seven sets of endocrine glands.

The seven centers—of which more will be discussed later—are often associated with areas of our body: at the base of our spine or sacrum, our genitals, solar plexus (just below the ribcage), chest, throat, brow, and crown of our head.

To distract the uninitiated the ancients disguised their knowledge of seven levels within us as myths and allegories, in mystical terminology, symbolic art, crafts, architecture, theater, dance, and as many of the symbols listed at the beginning of this chapter. Self-transformation was the significance of the mythic "rainbow bridge" of the Greek Iris, of the Norse Odin, and of Native Americans, linking earth and heaven, connecting our "lower" and "higher" natures. *We* are the seven-stringed lyre of Apollo, whose music is heard by both the gods and humans. The understanding of our sevenfold nature is hidden in the astrological symbolism of seven planets, the seven traditional vowels from alpha to omega, seven-headed snakes, seven-chambered caves of initiation, seven-stepped ladders to heaven and the seven gates of the "underworld" within us through which the Egyptian initiate had to pass toward liberation into light. The seven "liberal arts" of Medieval education—grammar, rhetoric, logic, arithmetic, geometry, music, astronomy—were intended as such a ladder to "liberate" us from a mundane life into the greater awareness and understanding worthy of the heavenly spheres. They are symbolized by the seven metals of the alchemical process that transmutes base lead (our familiar nature) to divine gold (spiritual adulthood). In seafaring societies our seven inner levels of experience were known as "the seven

Aztec depiction of seven "caves of initiation" within ourselves.

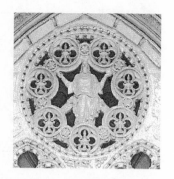

Lesser Rose Window of the Cathedral of St. John the Divine in New York City, world's largest cathedral (and still being built). Note its heptagonal design, one of the very few sevenfold Rose Windows in existence.

Odin's Rainbow Bridge to Valhalla.

Reconstructed Babylonian Ziggurat, model of both cosmos and self.

seas," across which none but the most worthy could travel. Our seven centers are symbolized by seven-story structures, mountains, and temples, from Chinese pagodas to the Potala, the former residence of the Dalai Lama, to the seven levels of Babylonian ziggurats, each level of which was colored to represent the seven visible celestial bodies, as were King Nebuchadnezzar's Hanging Gardens, one of the ancient world's Seven Wonders. The centers within us have been called the "seven daughters of Atlas," whom Zeus made into stars called the Pleiades, the seven sisters. In legend our centers were the "seven children of Pythagoras," three sons and four daughters, who may or may not have been historical, but who would be interpreted by a philosophical geometer as the triangle and square, the divine within us capping our mundane self.

Our sciences and mathematics are dominated by words having Greek roots, but we defer to India for terms describing details of our inner life, which were mapped long ago by Vedic priests. The Sanskrit term *chakra* ("wheel") is popu-

Egyptian god Djeheuti, or
Thoth, model for Hermes and
Mercury, holds staffs with
two twined serpents and
offers the "breath of life" to
the initiate Pharaoh.

Mesopotamian
cylinder seal

Roman caduceus

larly used to describe each of our seven centers because they
have been said to appear to the subtle sight of seers as
Catherine wheels, spinning fireworks.

Perhaps the most widespread and familiar of all ancient
esoteric symbols of the Self to survive the centuries is the
caduceus, a universal icon of medicine and healing. It was
sometimes depicted as a winged staff with two snakes twin-
ing up around it in the form of a double helix, crossing at
each of the five central *chakras*. It is startling to see this same
symbol depicted in so many different cultures. But its per-
vasiveness shouldn't surprise us since the discovery is a
human experience, identical for all regardless of the particular cultural symbols used to interpret and express it. The
caduceus was the staff of authority carried by the Olympian
messengers Hermes and Iris, and it became the medical
symbol of health when understood and adopted by Ascle-
pius, Hippocrates, and other famous healers.

The central staff of the caduceus is traditionally taken to
represent our spine, upon which the seven centers are often
shown attached like flowers on a stem. The path of the two
spiraling snakes spans the five middle centers from genital
to brow, symbolizing divine creative energies within us that
travel the spine in a subtly spiraling weave. Hermes and Iris
carry the caduceus as a symbol of their authority to travel
everywhere between heaven, earth, and the underworld,
the mythic world within us.

The caduceus, or spine, with its centers has been symbolized by seven-branched candelabras and sacred trees, by

the world axis, world mountain, maypoles, magic wands, swords, anchors, sacred rivers along seven cities, and by the staff of Moses, which became a serpent and saved "all who gaze upon it." Buddha's seven-year meditation under the bo tree (from the Sanskrit *Bo, Budh,* words signifying "awake") represents our inner journey through seven levels up the "tree" into enlightenment.

But these seven centers in us are not mystical tales, pagan superstition, or abstract concepts. They are familiar to everyone. We know them as our *psychological motivations,* the spectrum of the passions, desires, emotions, thoughts, and intuitions we experience, each with its own countless shades. The Pythagoreans called the number seven "the vehicle of life," alluding to these seven levels of motivation within us. Like each string of Apollo's lyre and each color of the rainbow, these centers vibrate at different rates and carry different qualities. We commonly associate these inner centers of subtle energy with parts of our body in phrases like

He who knows others is wise, but he who knows himself is enlightened.

—Lao-tzu

Human music is the harmony which may be known by any person who turns to contemplation of himself . . . It is this that join together the parts of the soul, and keeps the rational part united with the irrational.

—Gioseffe Zarlino
(1517–1590, Italian composer and musical theorist)

All that is in tune with Thee, O Universe, is in tune with me!

—Marcus Aurelius

Alchemical apparatus for purifying "metals"

The Hand of God, Celtic sculpture

Aztec worship of the double serpent

Sumerian sculpture

The central secret is, therefore, to know that the various human passions and feelings and emotions in the human heart are not wrong in themselves; only they have to be carefully controlled and given a higher and higher direction, until they attain the very highest condition of excellence.

—Vivekananda (1863–1902, Indian religious leader and teacher)

"thinking with his groin," "gut feeling," "have a heart," "all choked up," and "use your head." But what we actually *feel* at these places are universal energies streaming through us in rising and falling waves of differing lengths, intensities, and durations.

The two ends of our inner caduceus, sacrum and crown, forming its axis, act as two poles of a battery. Although their functions aren't apparent in ordinary human experience, they awaken in advanced stages of self-development. Notice how the two "snakes" do not reach the staff's ends but cross at five centers between the genital and brow. At the top their heads take the shape of our eyebrows.

Ancient twelve-tribe nations hid this inner symbolism in their astronomical and zodiacal myths. The seven visible celestial bodies listed in their traditional order (using familiar Roman names)—Sun, Moon, Mercury, Venus, Mars, Jupiter, Saturn—were associated with the crown, brow, throat, heart, solar plexus, genital, and sacrum. The illustration associates these seven centers with the seven planets in their "home" constellations. By examining their symbolism we uncover a beautiful allegory describing the birth and ascension of the divine power of consciousness within us.

One of the most misunderstood descriptions is the final chapter of the New Testament, the Revelation of Saint John the Divine. Written in mythopoeic symbolism, its descriptions of the transformative process startle the reader. In his

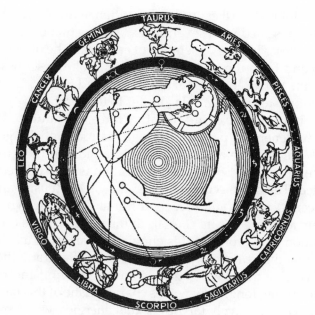

The Apocalyptic Zodiac

*M*en must be aware of
the wisdom and the
strength that is in them if
their understanding is to
be expanded.

—Vauvenargues (1715–1747,
French soldier, moralist)

*O*h, the difficulty of
fixing the attention of men
on the world within them!

—Samuel Taylor Coleridge
(1772–1834, English poet and
critic)

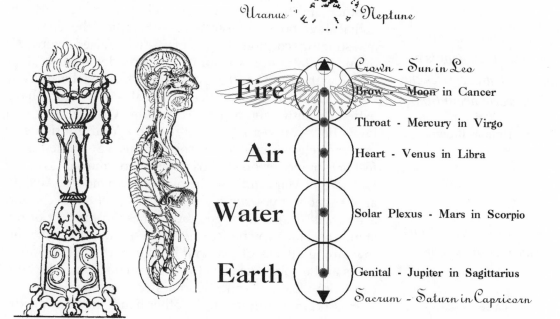

Pluto

Uranus   Neptune

Fire   Crown – Sun in Leo
            Brow – Moon in Cancer
            Throat - Mercury in Virgo

Air   Heart - Venus in Libra

Water   Solar Plexus - Mars in Scorpio

Earth   Genital - Jupiter in Sagittarius
            Sacrum – Saturn in Capricorn

Rejoice in being yourself a beautiful work of nature, and help yourself to further growth; that's the best thing.

—Moses Auerbach
(1812–1882, German novelist)

Choose always the way that seems right; however rough it may be. Practice will make it easy and pleasant.

—attributed to Pythagoras

There is no man alone, because every man is a microcosm, and carries the whole world about him.

—Sir Thomas Browne

The heart has its reasons which reason knows nothing of.

—Blaise Pascal

vision John describes a "book written within and on the backside, sealed with seven seals." We are that book.

He describes the "opening" of each seal. The first to open is second from the bottom, the genital. John describes it as "a white horse: and he that sat on him had a bow . . . and he went forth to conquer." He alludes to the symbol of "Jupiter in Sagittarius," the cosmic creating force, as "hunter." When creative energies passing through us focus at the genital level, we become aware of sexual passion and the roving hunter awakens. Sex in our civilization is both adored and reviled. The promiscuous Jupiter, the Greek Zeus, is always associated with a middle. So a human body drawn to fit within a square has its genitals at the middle. If we can pay attention to the force of this energy as it arises in us, distinguishing the *feeling* of the energy from its effects and the values we give them, we notice its power growing and receding in strong, long, slow waves characteristic of this passion. As we constructively tap our sexual energies, we access the power of the universal creating process.

The second center, associated with the solar plexus, is described by Saint John: "And there went out another horse that was red; and power was given to him that sat thereon to take peace from the earth, and that they should kill one another; and there was given unto him a great sword." In zodiacal symbolism we see "Mars in Scorpio," the red planet and stinging scorpion, the god of war with raised sword. We often see statues of warriors with Mars depicted on their armor, at the gut. The effects of energies focused through our solar plexus are often difficult to deal with, as they have extreme positive and negative attributes. On one hand, you know how you feel when you're angry. Your gut tightens; power swells within it. In the solar plexus we experience the feel of fear and terror, Phobos and Deimos, the names of the two sons of Mars and also the planet's two moons. Anxiety about school gives some children an upset stomach, and actors with stage fright get "butterflies" in the "pit of the stomach." Have you noticed how illness and weakening follow strong attacks of fear and outbursts of anger? Undue focus at the solar plexus is too often characterized by concern with what one can get, so desire, greed, and envy are also associated with it. On the other hand, the solar plexus is also the place to call upon for the noble characteristics of courage and endurance.

Whereas the solar plexus is concerned with what it can get, the heart, the third "seal," is concerned with what it can *give.* The astronomical symbolism is that of Venus, the goddess of love and beauty, wife of Mars and mother of Harmonia, at home in the constellation of Libra, the Balance Scale. Saint John's vision describes a horseman who "had a pair of balances in his hand." Our heart has its own way of knowing, of weighing events in ways transcending logic. The Egyptians called this "the intelligence of the heart." Picture someone or something you love and feel the warmth at the center of your chest. Sustain attention there, breathe with it, and feel the glowing warmth. Unless we succumb to thinking too much, we can feel the beauty of the heart unfolding and know its urge to give and love unconditionally.

On the other hand, the heart is all too often the unwilling container of repressed grief and anguish, sadness and disappointment. We can carry heartfelt emotions all through our lives even though we may bury them. Medical researchers are finding that lifelong repression of strong emotions without healthful release can lead to heart attacks. Sustained focus of attention at the heart center can be one of the most beautiful ways to experience life, but when the heart opens, those emotions, which have been most forcefully repressed, burst forth first. Some find this "dark night of the soul" difficult to face and shut down their emotions with logic and explanation, requiring a cleansing flood of tears to finally release the tension.

Among our spectrum of seven centers, motivations felt at the genital, gut, and heart are most familiar to people. Together with more rarefied motivations from the throat and brow centers they weave the Web of Athena, our personality pattern, which the Greeks personified as the goddess Psyche, our divine soul.

Plato wrote that to attend to the soul's development is the highest good we can do. The ancients were concerned with the soul's purification and transformation. As they saw it, the stream of energy called "desire" inevitably passes through us like a river. We cannot and should not try to cease

Love, who is most beautiful among the immortal gods, the melter of limbs, overwhelms in their hearts the intelligence and wise counsel of all gods and all men.

—Hesiod

When you have been compelled by circumstances to be disturbed in any manner, quickly return to yourself, and do not continue out of tune longer than the compulsion lasts. You will have increasing control over your own harmony by continually returning to it.

—Marcus Aurelius

There are a hundred and one arteries of the heart, one of them penetrates the crown of the head; moving upwards by it a man reaches the immortal.

—Khandogya Upanishad

Use the light that is in you to recover your natural clearness of sight.

—Lao-tzu

His aim should be to concentrate and simplify, and so to expand his being . . . and so to float upwards towards the divine fountain of being whose stream flows within him.

—Plotinus

its flow, but we *are* responsible for its qualities. For example, are we set on mindlessly acquiring more and more "things," or are our desires oriented toward wisdom and understanding? Do you see the difference? The ancients were concerned with such purification of each level by facing, releasing, and reorienting the energies of each center toward finer and more noble expression.

Experience with the centers above the heart involve advanced stages of meditation taught by a qualified instructor. Purified creative energies brought through the throat, brow, and crown call forth the opening of the "seventh seal" at our sacrum. The name *sacrum,* from the Latin for "sacred," is a curious name for the triangular bone embedded within the pelvis. The Freemasons symbolized the sacrum and spine as an upright shovel with which we dig the foundation of our "temple," our transformed self. It is the most profound and mysterious of the centers, and its lawful opening releases a transformative energy that results in a metamorphosis beyond what it is to be human. It is variously referred to as the crowning of the sun god, the arrival of the conqueror, and the return of the messiah, or avatar, crowned with seven rays and surrounded by twelve stars. We miss the point if we misinterpret the ancient myths by literalizing this divine emergence as only having once occurred to someone else, outside our own being. The ancients taught that it is *we* who have this possibility of liberation.

Continuing the metaphor of the prism and the spectrum, when all seven centers have been purified, their colors recombine into whole white sunlight and are offered to their source "above" or deep within us. Purification is a major

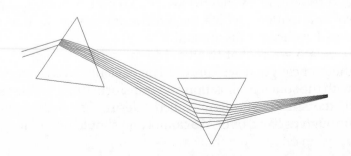

concern of fairy tales. It was only after Snow White cleaned and organized the home of the seven dwarfs that they were able to come to her rescue, allowing the prince to awaken her with his kiss and then take her to live in his father's divine castle.

The details of this process were the most closely guarded secrets of the ancient world's religions and mystery schools. Like people of all eras and places, we can discover the archetypal source of our motivations for oneself only by direct experience. And we must start exactly where we are, simply by becoming more aware of our motivations and by perceiving our inner life as dynamic waves of passing energy. In ancient times, becoming aware of oneself as an energy system was symbolized as finding one's way onto the rainbow bridge linking "earth" with "heaven." Life was likened to the song played on Apollo's lyre; one tuned and ascended this musical scale of Apollo's lyre within oneself. The ancients sought to tune or purify themselves so that the unseen divinity deep within them could better play upon their instrument in the world.

To the ancients, the task of constructing the universe meant reconstructing themselves. In democratic societies of today we are not constrained as in the past. Timeless models of ourselves and tools for self-awareness are becoming understood again. It is unnecessary to hide this knowledge of our inner constitution. It is more important to understand the sources of our motivations and to transform them. Once we catch on, we travel in seven-league boots.

Native American mask, Statue of Liberty, and Egyptian Goddess Seshat ("Seven") each represent an enlightened being crowned with seven rays.

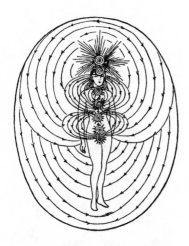

Each of us is a tripart whole,
an individualized field of
interpenetrating light, energy,
and living matter, or Higher
Self, psyche, and body.

The seven-stage transformation of our elusive inner life, the goal of ancient seers, philosophers and initiates around the world, does not graft miraculous powers onto our familiar Self; it changes us from our core. After all, a butterfly is more than just a caterpillar with wings. The process transforms our very mode of perception: we see ourselves directly, not through reflected pictures of ourselves and the world but as the very energy that motivates those pictures. It's no wonder that the sevenfold processes of nature and mathematics had to be made metaphorical through mythology and religion, fairy tales, and architecture. Seven is an unborn virgin, eternally elusive when chased by words or the geometer's tools. But, to our surprise, it is the Heptad's role in our inner awareness that is also doing the chasing.

Untwisting all the chains that tie the hidden soul of harmony.

—John Milton (1608–1674, English poet)

CHAPTER EIGHT

OCTAD

Periodic Renewal

Change has an absolute limit:
This produces two modes;
The two modes produce four forms,
The four forms produce eight trigrams;
The eight trigrams determine fortune and misfortune.
> —*Confucius (commentary on the* I Ching)

Now this, monks, is the noble truth of the way that leads to the
cessation of sorrow: this is the noble Eightfold Way; namely, right
views, right intention, right speech, right action, right livelihood,
right effort, right mindfulness, right concentration.
> —*Buddha*

Mathematics takes us still further from what is human into the
region of absolute necessity, to which not only the actual world,
but every possible world, must conform.
> —*Bertrand Russell (1872–1970, British mathematician*
> *and philosopher)*

In the uncertain hour before the morning
Near the ending of interminable night
At the recurrent end of the unending.
> —*T. S. Eliot (1888–1965, British poet and critic)*

You can't fold a piece of paper in half eight times.
> —*Math folklore*

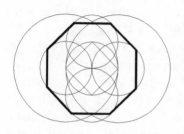

The Birth of the Octagon.

BEYOND THE EIGHT BALL

Passing beyond elusive virgin number seven, we leap to eight and find ourselves in a quite different place within the Decad. Eight is promiscuous, having more divisors among the Decad than any other number; it is divisible by one, two, and four ($8 = 1 \times 2 \times 4$). Its divisors tell us that its roots are clearly in the Monad, Dyad, and Tetrad. The ancients called eight "justice" and "evenly even" because it can be continually halved all the way to unity (unlike an "oddly even" number like six, which cannot reach unity by halving).

The Indo-European names for "eight" emphasize it as a "doubling of four," most deriving from the Sanskrit *o-cata-srah* ("twice four"), which became *okta* in Greek and *octo* in Latin. These references to one, two, and four tell us that the archetype of eight, called Octad, weaves together the principles of Monad's unity, expansion, and cycles, Dyad's polarity, and Tetrad's materialization, that is, material forms precipitate from their archetypes through polarity and pulsing cycles.

Within the Decad only one and eight are composed of three identical divisors ($1 = 1 \times 1 \times 1$ and $8 = 2 \times 2 \times 2$), and so they project into three-dimensional space as cubes having sides of one and two, with volumes of one and eight. Every cube has eight corners as if to emphasize the mutual relationship between the number eight and this form. The cube's dual, the octahedron, has eight faces dividing space equally in eight directions.

Square faces link the cube and the Octad with the Tetrad's symbolism as "earth," manifest form, the crystalline phase of matter, and physical stability. This is why the number eight, like the number four, is often associated with the Great Mother Goddess in her nourishing aspect.

Folk references to "eight" are not particularly common. We speak of the "figure eight" in ice skating and of being "behind the eight-ball." Other than the word "eight," what's their connection? The figure eight represents a polar cycle perpetually swinging between extremes and crossing itself in the middle as the diagonals of a square. Turned sideways, the figure eight is the mathematician's glyph for the infinite.

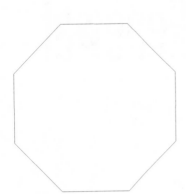

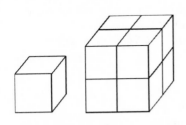

1 x 8

2 x 4

4 x 2

8 x 1

One and eight are the volumes of the only two cubes within the Decad.

The significance of the phrase "behind the eight ball" seems more obscure unless we realize that the black billiard ball numbered "eight" is like the crossing point in the figure eight. It is the "oddball" at the exact middle of the fifteen balls, with seven lower-number solid-color balls below and seven striped balls with higher numbers above. To be "behind the eight ball" is to find oneself in the middle of a dilemma with (seven) elusive solutions on each side.

The seventeenth-century Spanish gold coins known as "reales," and favored by pirates, were called "pieces of eight" because each coin was perforated into eight pie-shaped sections and could be broken into fractions of their full value. Two "pieces of eight," one quarter of the coin, is the source of the use of the American slang term "two bits" to signify a "quarter" dollar.

The number eight is particularly emphasized in the mythology and religious symbolism of the Orient. The Chinese revere the Eight Immortal saints, eight symbols of a scholar, and eight directions of the wind. Buddhism is known by eight auspicious emblems and particularly by the Eightfold Path leading to enlightenment. To understand the significance of the Octad in cultural expressions we must examine its appearance in geometry, nature, and art.

Notice how the octagonal shape is easily identified from a distance.

The Eight Immortals of Chinese myth, who gained their stature by studying nature's secrets, wish us good luck.

I am One that transforms into Two,
I am Two that transforms into Four,
I am Four that transforms into Eight.
After this I am One again.

—Egyptian (Hermopolitan) creation myth

THE BIRTH OF THE OCTAGON

The geometric construction of the regular octagon calls upon eight's divisors one, two, and four, numbers of the Monad, Dyad, and Tetrad. This tells us that the construction involves the circle, *vesica piscis*, and square. The way their principles come together as the Octad is best understood by facilitating its geometric birth through the *vesica piscis* as an octagon.

(1) First, construct a *vesica piscis* and a square within it. This square is the construction's "seed" or guiding form. (2–5) With the compass open to the radius of the inner circle, put its point on each of the square's four corners and turn a circle centered at each corner. (6) Using the straightedge, extend the original square's four sides until they completely cross the four circles created at the corners. (7) Connect the eight points where the extended lines intersect the circumferences of the two large circles to (8) make an octagon. (9) Use the straightedge to connect every *other* corner of the octagon to create two interlaced squares within it. (10) Also extend each of the octagon's sides to form interlaced squares around it. The octagon and interlaced, or double, square are the basic structures necessary for examining the Octad's principles.

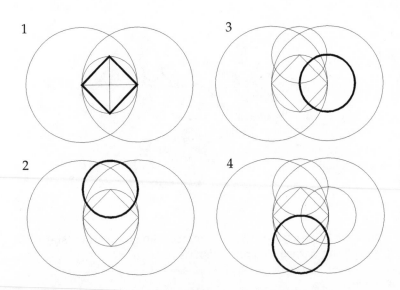

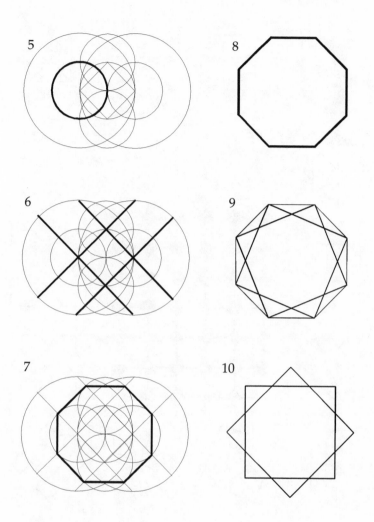

EXPLORE THE OCTAGON'S GEOMETRY

Use octagons you construct as starting frames to explore the form's internal structures and patterns. Using the straight-edge, connect corners and secondary crossing points to reproduce some of the patterns shown, or your own.

You may wish to make the constructions *audible* by using wood, nails, and a guitar string, piano wire, or nylon fishing line to configure the lines of the construction. Plucking the "line," or string, between different points will enable you to hear the octagon's internal proportions in relation to one another. You will discover some tones that are identical with

others but higher or lower in pitch. The octagon's geometry is similar to that of the square, built on the "root-two ratio" (1.414. . . to 1), which you will hear.

Octagons alone will not tessellate, or fill flat space, without leaving gaps. But they will do so when combined with small squares. To construct this familiar pattern seen in tiles, wallpaper, fabric, and elsewhere, construct or doodle an array of circles aligned in square-pattern close-packing (as opposed to hexagonal close-packing). Fill the gaps with small, tilted squares, connect their corners, and watch the circles transform into octagons.

DISCOVER THE OCTAD'S PATTERN IN LIVING FORMS

We've been seeing the most widely varied geometries occur in the living structures of the sea. Along with circles, spheres, triangles, squares, pentagons, hexagons, and the five Platonic volumes, the geometry of the Octad can be found in the structures of such creatures as jellyfishes and "glass" sponges. Place tracing paper over each image, and connect the corners and crossing points of the interlaced squares and octagons that enclose each form to discover the creatures' underlying geometries.

Seen from below, many jellyfishes and "glass" sponges display octagonal structure.

Male maple flower

THE BREATH OF THE COMPASSIONATE

Islamic tradition holds that there are one hundred names of God but only ninety-nine are knowable and speakable, and they are called the 99 Beautiful Names of Allah. The highest pronounceable name is "The Compassionate." From its infinite goodness the Compassionate exhales and inhales. Through the polar cycle of divine breath the universe is periodically created, maintained, dissolved, and renewed.

Islamic design, in which the depiction of animals or humans is forbidden, symbolizes this process through geometry. The pattern known as the "Breath of the Compassionate" appears throughout Islamic art and architecture, in wall tiles, ornamentation, doors, screens, rugs, manuscript illumination, and elsewhere. In constructing this pattern the geometer calls upon the three factors of eight (one, two, and four), weaving the principles of the Monad, Dyad, and Tetrad, expressing cycle, polarity, and material form.

Begin with a central point through which the unknowable emerges. Turn a circle, duplicate it as a *vesica piscis*, and (1) construct a square (symbol of material form in the four states of *mater*). The archetypes are becoming visible through the creating process. (2) The square evolves to become an octagon and interlaced squares containing both elements of the full breath. (3) When the eight corners of the interlaced squares are unfolded, pointing *outward*, we create the symbol of cosmic expansion, one exhalation of the Breath of the Compassionate. (4) A second construction turns the corners *inward* to depict contraction and inhalation. (5) These "breaths" will "tessellate," or fill the plane in all directions, leaving no gaps, just as Divine Compassion fills the universe without gaps, creating a geometric motif seen throughout Islamic design as a reminder of divine presence. This motif tells us that the process by which the archetypal patterns crystallize into material configurations is a polar process as simple and ephemeral as breathing. And so it is in the basic positive and negative charges of the atoms that configure the universal patterns. We can share the compassion that creates and fills the cosmos by feeling it along with each of our own breaths.

God created the universe through the Breath of the Compassionate.

—Mohammed (c. 570–632)

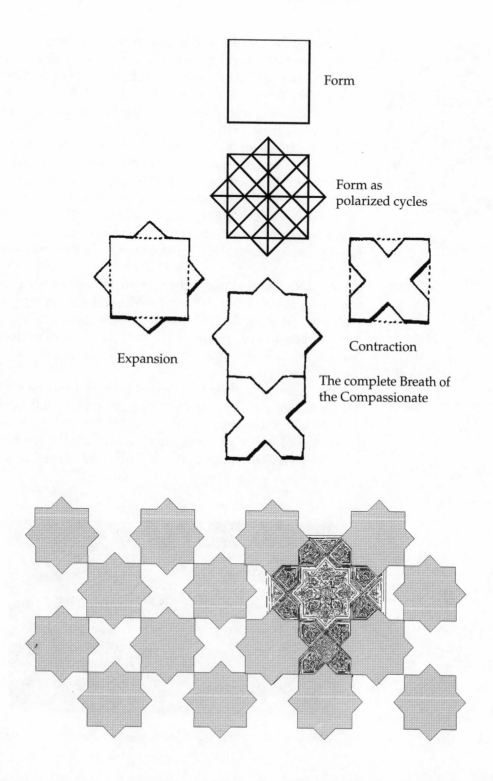

Form

Form as
polarized cycles

Expansion

Contraction

The complete Breath of
the Compassionate

Islamic pattern

More than just an ornamental motif, the Breath of the Compassionate is a cosmological model symbolizing the interplay of polarities that manifest form. Islamic tradition, whose originators voraciously translated all the ancient Greek scientific, mathematical, and medical texts they could find, tells us that, on one level, its eightfold pattern refers to Aristotle's diagram of the four elements, or modern states of matter, or *mater* (earth = solid, water = liquid, air = gas, fire = electronic plasma), crossing with four qualities (cold = contraction, hot = expansion, wet = dissolution, dry = crystallization). On this eightfold loom the archetypes precipitate. Each geometric inhalation and exhalation of the Compassionate represents this eightfold weave of the configured cosmos.

Aristotle learned this symbolism from his teacher Plato, who had studied the Pythagorean teachings. Pythagoras had learned it in Egypt, where it took the form of the primeval ogdoad of Hermopolis, the four pairs of unmanifest gods and goddesses responsible for giving birth to the manifest universe.

Eightfold geometry is an ancient symbol for the Great Mother Goddess as nourisher. Her substance clothes the archetypal patterns by crossing the four elements with four (two pairs of) opposite "qualities." In this role she has often been depicted mythologically as an eight-legged spider

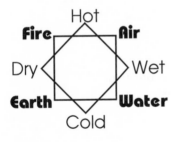

Aristotle's depiction of the four "elements" and the qualities that form them.

Sea spider

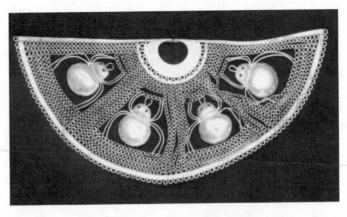

Peruvian gold nose ring. Half a circle contains four spiders.

(from the word "spinner"). In Native American mythology, for example, the Grandmother Spider weaves the web of matter. In modern terms this would be done by intricately crossing polarities, electrons and protons, the warp and woof of each atomic knot. The Greek word *Kosmos,* which literally signifies "embroidery," and many of the great goddesses of myth (the Egyptian Neith, Nutet and Isis, the Greek Athena and the Three Fates, and the Mayan Ixchel) are spinstresses, weavers, and embroiderers, manifesting the world while spinning the threads of fate for humans and the entire cosmos.

Spiders are not insects, although both are "arthropods" (from the Greek for "jointed feet"). Insects have three body sections, six legs, and two antennae, while spiders have two sections, eight legs, and no antennae. Spiders are arachnids, named for the Greek maiden Arachne, who impudently

Native American carved shell depicting Grandmother Spider.

Plate from Panama

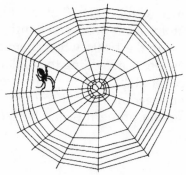

*T*he spider's touch, how
exquisitely fine!
Feels at each thread, and
lives along the line.

—Alexander Pope

Eight legs of the spider, eight tentacles of the octopus, and eight directions of a Hindu Kali Yantra meditation mandala all symbolize the ensnaring, devouring, terrible aspect of the Great Mother Goddess.

challenged the goddess Athena, inventor of weaving, to an embroidery contest and was turned into a spider for her arrogance.

In her terrible aspect, the spider goddess is the ensnaring feminine dangling from the heavens to trap the unwary in her web. Foreboding doom, she ultimately provides her catch with a fate of spiritual transformation and renewal. Numerous spider emblems have been found throughout the world, from Crete to Germany, Mexico, and Peru. The eightfold terrible transformer has also been represented as an octopus with its eight ensnaring arms, which spiral to reinforce it as a symbol of transformation.

THE PERIODIC TABLE OF THE ELEMENTS

Scientists have found a modern analogy for the eightfold weave of the cosmic spider web in the Periodic Table of Elements, the "map" of all ninety-two naturally occurring atoms from hydrogen to uranium.

In the 1860s, when chemists began grouping the known elements according to similar physical properties, they discovered that the properties, repeat in periodic cycles. A table emerged of vertical columns called "families," which have similar properties, and horizontal rows marking gradations

I	II												III	IV	V	VI	VII	VIII
H				Transition elements														He
Li	Be												B	C	N	O	F	Ne
Na	Mg	III	IV	V	VI	VII	VIII			I	II		Al	Si	P	S	Cl	Ar
K	Ca	Sc	Ti	V	Cr	Mn	Fe	Co	Ni	Cu	Zn		Ga	Ge	As	Se	Br	Kr
Rb	Sr	Y	Zr	Nb	Mo	Tc	Ru	Rh	Pd	Ag	Cd		In	Sn	Sb	Te	I	Xe
Cs	Ba	La	Hf	Ta	W	Re	Os	Ir	Pt	Au	Hg		Tl	Pb	Bi	Po	At	Rn
Fr	Ra	Ac																

La	Ce	Pr	Nd	Pm	Sm	Eu	Gd	Tb	Dy	Ho	Er	Tm	Yb	Lu
Ac	Th	Pa	U	Np	Pu	Am	Cm	Bk	Cf	Es	Fm	Md	No	Lr

Periodic Table of Elements

of those properties, each element having one more electron than the element to its left, thus forming "spectrums" of matter across the table. Each element has properties similar to those directly above and below it, a little different from those to the left and right. The chemists discovered eight main groups, or types, of elements.

Scientists now theorize that atoms and the world's atomic configurations are not "things" but whirlpools of *energy*, swirling clouds of negatively charged electrons surrounding a positively charged core, or nucleus, of protons and neutrons. The reason the atoms' properties recur in periodic cycles of eight, as the scientists eventually discovered, is that elements in the same column have the *same number of electrons*, between one and eight, in the outermost periphery of their orbits. The "group" number at the top of each column tells the number of outer electrons. When chemical elements combine they form molecular architectures determined by the numerical and geometric patterns of the outermost electron rings, or "shells." Atoms strive to combine with other elements that will give them *eight* electrons in their outermost orbit, making them stable and satisfied elements with a "full shell." Atoms at the extreme right column of the table, known as the inert, or noble, gases like neon and argon, already have a full outer shell of eight electrons and are complete unto themselves. They are reluctant to combine with other elements. When the outer shell is full a new row, or "period," of elements begins again with one electron. Each consecutive atom has similar physical, chemical, electronic, optical, or crystalline properties similar to those above.

For example, consider the elements oxygen and sulfur, directly below it. Both have six electrons in their outer shell and similar chemical properties. They will each gladly combine with two atoms of hydrogen, whose single electrons will complete the outer shell with eight electrons. Oxygen and hydrogen combine to form H_2O or water; sulfur and hydrogen form H_2S, or hydrogen sulfide. Eight sulfur atoms by themselves, in fact, will share their outer electrons to combine in an octagonal ring (S_8), each atom now having a full shell of eight electrons. Modern chemistry is discovering molecular, atomic, and subatomic structures strikingly sim-

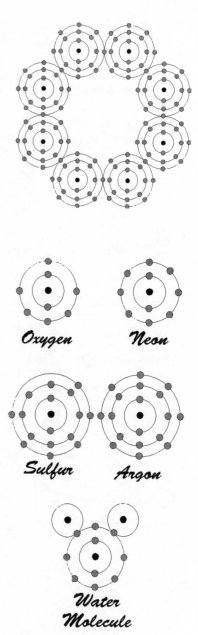

Oxygen   Neon

Sulfur   Argon

Water
Molecule

ilar to the geometric patterns of Islamic art and architecture.

The Periodic Table of Elements is a modern cosmological model, a contemporary depiction of the Breath of the Compassionate and the cosmic web of the Grandmother Spider. It is the web of the world, weaving the principles of the Monad, Dyad, and Tetrad. That is, it depicts a universe woven of polarities, positive and negative charges of subatomic matter, and recurring eight-step cycles to manifest all matter. Each horizontal row of the Periodic Table represents another "octave," or eighth step, up the "scales." When each row's final chemical "note" is reached and the outermost electron shell is filled with eight electrons, the next row, or "octave," begins. The entire Periodic Table, the complete list of known types of matter, spans seven "octaves," the number appropriate for the full spectrum of material configurations.

Opinion says hot and cold, but the reality is atoms and empty space.

—Democritus (c. 460–370 B.C., Greek philosopher)

RESONANCE: THE SAME BUT DIFFERENT

In a sense the Periodic Table is like a piano keyboard, each atom a chemical "note" along an ongoing scale. The materialized forms the atoms weave are like visible music. Each eighth atomic group and eighth white key on the piano comprises one turn of its spiral, repeating the tone and chemical properties at each rung. Whether musical notes are repeating their tone or elements are repeating their physical properties, each "octave"-step restores the fundamental base in a process of self-renewal, a periodic recurrence, a return, but to a corresponding rung of the spiral.

The idea of periodic recurrence is at the heart of "resonance," or resounding, the same note an octave apart. We honor the principle of resonance when we celebrate holidays and anniversaries on the same day each year. A simple experiment with a guitar will demonstrate how "separate" physical objects are linked through this principle.

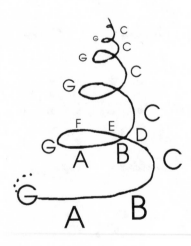

Pluck different strings, and listen to their notes. Now crease a very small piece of paper and hang it on one of the strings. Pluck the other string. Nothing happens to the paper. Now tune the strings to sound exactly the same note and pluck the other string again. The paper jumps as its string "resonates," or vibrates, in tune with the plucked string. The strings are connected through an energy link.

Now press your finger at the exact center of the string without paper and pluck *one half*. The note produced is the same as the one sounded by the whole string, but it is an octave higher. The two strings again resonate and vibrate together and the paper jumps again. Although physically separate, the strings vibrate at the same rate and align through space by means of their energy link.

Resonance occurs between plucked strings as they are continually halved or doubled in length, creating the same tone in higher or lower octaves. Similar musical notes across any number of octaves will resonate with each other just as the properties of atoms resonate with those listed above and below them in the Periodic Table.

Everything in the universe has its natural vibration frequency. Just tap and listen. All similarly vibrating objects of the material and energetic universe are potentially linked by resonance. You may have noticed this while driving a car. As the car accelerates, its pistons oscillate, or vibrate, at different rates. At a certain speed, or rate of vibration, the windows may rattle. At a higher speed, the windows stop but the glove compartment door rattles. Then the sound recedes and perhaps the door handles begin to rattle. As the car's vibratory "note" changes, those loose objects that resonate with that particular "note" will absorb its energy and vibrate or rattle with it.

The Octad's principle of resonance goes far beyond a rattling car. The discovery of resonance in oscillating electronic circuits led to the idea of "tuning" a radio receiver to different stations instead of fixing it to receive only one station. That is, the radio receiver resonates in synchronization with the broadcast waves, whose motions make the paper cone of the speaker jump and sing, reminding us of the way the paper jumped from the guitar string.

On the average, the natural speaking voices of males and females are one octave apart. This fact confirms what we

So far as it goes, a small thing may give analogy of great things, and show the tracks of knowledge.

—Lucretius (c. 99–55 B.C., Roman poet and philosopher)

already know: men and women are "the same but different," equal without being identical.

Through the process of continual halving or doubling we introduce the Dyad and polarity to expand the octave infinitely. The eighth step is the return, the periodic renewal to the source, to the Self. We arrive at a higher or lower rung of the spiral, the same but different.

The number eight itself is the result of a triple-doubling from unity, the Monad: $1 \rightarrow 2 \rightarrow 4 \rightarrow 8$. We know that our body grows from the time sperm and egg unite to form a whole cell, which doubles to become two cells, then four, eight, and so on to form tissue, organs, and systems. But the doubling from one cell to two in the process of "mitosis" (from the Greek *mitos* signifying "thread," referring to the threadlike appearance of DNA chromosomes in the cell's nucleus) is a process of eight seamless stages, as if along the tones of a seven-note musical scale that renews itself upon reaching the eighth step, where it becomes two.

Wherever we see an ongoing cyclic process involving polarity or doubling and resulting in material manifestation, we can expect an eightness and the principles of the Octad.

There is a curious mathematical relationship between the numbers seven and eight, the musical scale and octave, and the process of doubling. Use a calculator to divide 1 by 49 (=7x7), that is, 1/49. The answer (.020408163264128256...) is an endless decimal built on the process of doubling.

Cell division. One cell becomes two along an octave of eight stages in the process of *mitosis* (Greek for "thread formation").

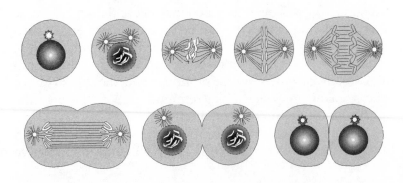

It starts with 02, then 04, then 08, then 16, 32, 64, 128, ... The mystery of the creating process jumps out at us wherever we look.

We see this occur at the shore in the breaking of a foaming spiral wave. Incoming ocean waves will break when the distance to the bottom below the wave equals exactly half the surface distance between two waves, their "wavelength" (see chapter five).

A simple doubling relationship occurs within the structure of the human body. Your thumb, wrist, neck, and waist form a series of doubling "octaves." Wrap a length of string around each to verify that:

* twice around the base of your thumb = once around your wrist

* twice around wrist = once around your neck

* twice around neck = once around your waist

Few of us will show precise results, but with many people the average relationship will be close.

Resonance explains why we can sometimes see through a "solid" object like glass. Think about it. How many solid objects can you actually see through? When light, either coming directly from a source or reflected from an object, strikes the surface of glass, its molecules of silicon dioxide [SiO_2] absorb the energy and oscillate with it, ringing like microscopic bells, vibrating in the same range as visible light colors. The tune is perfect, and nearby "molecular bells" absorb the energy through resonance and ring with the same note. Eventually, the vibrations reach the other side and emerge as light nearly identical with their entry vibrations. Thus we "see" the object on the other side. In actuality we are seeing light that has resonated through matter. As $E=mc^2$ tells us, energy (E) is the link between matter or mass (m) and light (c).

We use the word "transparent" to describe objects like glass whose molecules resonate with light completely. When only *some* of the molecules in a substance resonate or are "tuned" slightly differently, only some of the light resonates through and we see, "as through a glass darkly," that the material is "translucent." When *none* of the molecules in an object "rattle" in tune with the light, the object is "opaque."

Resonance in the world has long been recognized. Formalized in the Middle Ages as the Hermetic Doctrine of Correspondences and the Law of Analogy of Similars, it is still the basis of primitive "sympathetic magic" around the world. Resonance is behind the symbolism of the name of the Greek god of harmony "Apollo," which signifies "moving together." It was also at the heart of ancient therapeutic music, wherein the tones of the tuned lyre and pure voice struck chords corresponding to the tones of the elements of the psyche and body.

We experience resonance on a very personal level in our rapport with the people we know and meet. Consider that our inner world is an energy system. Each urge, desire, emotion, thought, and intuition represents a different wave band of subtle energy. Each of us unconsciously broadcasts the energy of our inner life and receives only that with which we are in tune.

Have you ever been in a crowd of people and felt your mood swing drastically? We resonate with the prevailing subtle "music" if we have cultivated that "tone," or quality, within ourselves. The ancients knew that unless we purify and tune our inner life and center ourselves at as high a level as we can, we will resonate with whatever passions are blowing in the wind. This is the root of much unconscious suffering, not knowing the sources that motivate and shape the pliable human psyche.

Our outer experiences only resonate with our inner rattle. If we don't like the connections we've formed we cannot break them, but we can transform ourselves so as not to resonate with them. It's useless to blame anyone or anything outside ourselves. We carry our inner state with us and will find the same types of friends and relationships wherever we go. To change what we encounter in the world we change the levels and qualities to which we attune.

In myth, legend, and religious symbolism, the leap to an eighth step is traditionally associated with spiritual elevation and deliverance, or salvation. Thus, we often see in

*T*hose in whom there is no guile do not remain in the seventh, the place of rest, but are promoted to the heritage of the divine beneficence which is the eighth grade.

—Saint Clement of Alexandria (c. 150–215)

architecture a square base below an octagon supporting a spherical dome, the octagon symbolizing the transition between the square and sphere, earth and heaven. In the alphanumeric system of the early Christians, based on the Greek language in which the New Testament was written, the letters of the name of Jesus, Ἰησοῦς, add up to 888, emphasizing the transcendent symbolism of the octave.

HARMONY OF THE COSMIC BREATH

In our examination of expressions of the Octad we find the ancient Chinese oracular system known in the West as the *I Ching*, or *Book of Changes*. Regarded by modern science as superstition, it is actually an interesting cosmological model. In many ways the *I Ching* is like the Periodic Table of Elements; both systems are based on various combinations of polarity and periodic renewal. But instead of looking at "things" and structures like "atoms" and "molecules" the *I Ching* depicts the cosmos as dynamic events, as distinct

stages of transformation, the processes that link "things." This process was called the Tao, the way of natural law. The *I Ching* was used by ancient Chinese sages to explain all processes of growth and transformation, from plant and animal life cycles to the movements of the heavens and the procedure of events in our lives.

In Chinese myth and legend the unknowable Absolute, known as Tai Chi, the "Great Extreme" or "Grand Ultimate" (literally "ridgepole," the world axis), issues the Monad, Tai Yi, the "Great Unity," symbolized by a circle. This One, then, produces the complementary powers of the Dyad, the *yang* and *yin*, polarities of the world: positive-negative, light-dark, male-female, odd-even, advance-withdraw, active-receptive, proton-electron, aggressive-yielding, involvement-detachment, motion-tranquillity, and so on (see chapter two). In its circular form, each part contains a small piece of its opposite within itself, causing the parts to endlessly chase each other. If we symbolize the *yang* as a solid line and *yin* as a broken line, their combinations taken two at a time give us exactly four possibilities. Taken as "trigrams," three at a time, we find exactly eight possible combinations of *yang* and *yin*. In the process of transformation,

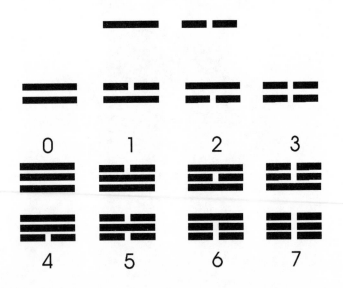

Yang and Yin taken two at a time produce four possibilities. Taken three at a time they yield eight possibilities.

each of the three levels comprising the trigrams can be either positive or negative, *yang* or *yin*. Each level is capable of change, and as it progresses through its eight transitional states the "charge" of each position alternates. Look closely at the subtle transformative stages of the lines as they proceed from three solid to three broken lines. With your eyes you can feel their rocking rhythm. Modern mathematicians will recognize this logically expanding pattern as the binary, or "base two," sequence counting from zero to seven, the code with which modern computer software is written, suited to electricity's only two possibilities of being "on" or "off." (Computer hackers measure a computer's memory in terms of these eight combinations as eight "bits" to the "byte," which build to kilobytes and megabytes). By examining the occurrences and combinations of active and passive forces in their various positions and relationships within each "trigram," the sages realized how the rhythms of polarity give birth to all possibilities.

According to tradition, in 2825 B.C. the legendary first ruler of China, Fu Hsi, came across a tortoise shell. In a flash of insight upon examining its pattern, he understood the various ways in which the cosmic polarities interplay in any situation. He composed his findings as the *I Ching*, which was modified seventeen centuries later by King Wen and six centuries after that by Confucius. The text that has come down to us is the result of their combined understanding and commentary.

If we look at a tortoise shell we find tessellating hexagons arranged in a six-around-one pattern. The principles of the hexagonal shape tell us that structure, strength and material efficiency are at work—an obvious advantage for a tortoise shell. But on further examination if we take two hexagons as the center we see that they are surrounded by *eight* other hexagons. Fu Hsi saw in the two central hexagons the opposite principles of *yang* and *yin*. In the wrinkles within the surrounding hexagons he saw their eight possible "trigrams." The diagram of the eight trigrams around the central symbol of *yin-yang* is known as the Pa Kua ("eight diagrams"). There are actually two traditional arrangements of the eight around the center. The older, Fu Hsi's original (depicted here), is known as the Former

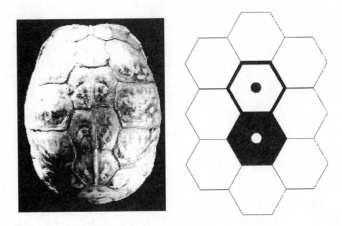

Unceasingly contemplate the generation of all things through change, and accustom thyself to the thought that the nature of the universe delights above all in changing the things that exist and making new ones of the same pattern. For everything that exists is the seed of that which shall come from it.

—Marcus Aurelius

Heaven Sequence and is said to depict the ideal pattern of transformations of the cosmos. Another arrangement of the same eight known as the Later Heaven Sequence is used by geomancers to interpret transformations on earth, enabling them to determine the harmonious placement and alignment for cities, towns, temples, homes, and tombs in relation to the landscape and its terrestrial energies.

Notice how each solid and broken line forms a symmetry with its corresponding place across the central *yin-yang* symbol.

Fu Hsi interpreted the different combinations and placements of *yang* and *yin* as a cycle of the stages in any process. The cycle begins with the three *yang* lines at the top, representing "heaven," or the most active state. The process proceeds through eight stages, not *around* the circle but in a "figure eight," crossing the center between stages numbered three and four, where it continues around and reaches the bottom, the three *yin* lines representing the "earth," the most passive and receptive state. Being a continuous process it crosses up the center again and returns to the most active stage of the cycle, pure *yang*. The eight stages are an expression of the archetypal Octad, the result of the play of principles of Monad, Dyad, and Tetrad. The ancient names of the eight trigrams and their symbolism are said to reveal the essential structure of every cycle found in nature. Read them in sequence from 0 to 7 and then 7 to 0. Try to sense the flow of their cycle.

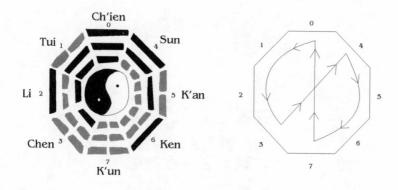

0 ☰ Ch'ien   Heaven, creative, origin, immobile, strong, firm

1 ☱ Tui   Attraction, achievement, joy, satisfaction

2 ☲ Li   Awareness, separation, beauty

3 ☳ Chen   Action, movement, development

4 ☴ Sun   Following, penetration

5 ☵ K'an   Danger, peril

6 ☶ Ken   Stop, rest

7 ☷ K'un   Earth, receptive, docile, flexible

To examine the cosmic transformations in greater detail, the sages combined the eight trigrams in pairs, one over another forming six lines called "hexagrams" (not to be confused with the six-pointed star). The resulting sixty-four combinations reveal the continuous stages in the cosmic creating process flowing from all *yang* to all *yin* and back around. They are traditionally arranged as an 8 × 8 square and also as a continuous circle. Each hexagram represents a triad: *yang, yin,* and the harmonious balance between them.

The sixty-four transformations can be arranged as a heavenly circle, or square earth.

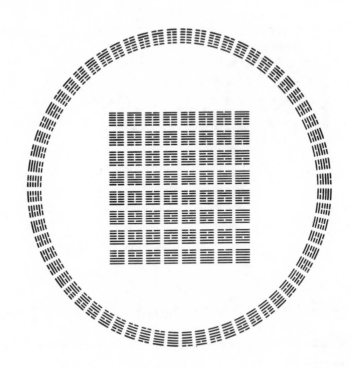

The sixty-four hexagrams form the archetypal evolution of events within the continuous flow of all possibilities.

As an example, the hexagram symbolizing "gradual progress" is made of the trigram for "rest," below the trigram for "penetration." The relative positions of the trigrams modify their significance. Inverted, "rest" capping "penetration" indicates "decay."

The ancient sages referred to the pulsing rhythm of polarities within each hexagram as the "harmony of the cosmic breath." Like each "exploded" and "imploded" octagon of the Islamic "Breath of the Compassionate" the hexagrams reveal the Octad's cycle of polar possibilities.

While the name *I Ching* is usually translated as the *Book of Changes*, the word "I" can be taken not only as "change" but as "easy," referring to the natural path along which the universe flows and which the sage knows as the path of the cosmic creating process. At any given instant the universe may be characterized by one of these sixty-four stages. The *yin* and *yang* are polar opposites, and each moment in time

Penetration

Rest

"Gradual
Progress"

Rest

Penetration

"Decay"

is the third element, the Triad, in which they merge and harmonize, for each moment is harmonious within a great order. The act of tossing and observing three coins, three tortoise shells, or the groupings of forty-nine yarrow stalks represents the *yin* and *yang* possibilities of the moment they are tossed. Each hexagram reveals the cosmic polarities at work, a "picture" of the moment. Thus the moment, the hexagram, and the question in mind at the time of tossing are synchronous. In ancient times no interpretive book like the *I Ching* was necessary because the sages could decipher the hexagrams by examining the lines and their positions directly. By interpreting each moment's hexagram the sages could take the proper course of action warranted by that moment. The *I Ching* made the oracle more widely available.

In their cosmological models all cultures recognize the interplay of polarities as the most fundamental aspect of the creating process. This is true in nature as well as in the symbolic constructions of the geometer. Hindu Ayurvedic medicine recognizes six basic "tastes": sweet, sour, bitter, salty, pungent, and astringent. Combinations of their presence or absence in food form sixty-four possible tastes we can recognize. Modern science is finding this arrangement in fields as diverse as computer design and biotechnology, where investigations into the genetic code have revealed DNA to be composed of sixty-four six-part "codons," or genetic "words."

The cosmic structure-function-order remains the same through time's transformations, whether interpreted by ancient symbology or modern terminology.

> *The I Ching does not offer itself with proofs and results; it does not vaunt itself, nor is it easy to approach. Like a part of nature, it waits until it is discovered.*
>
> —Carl Jung

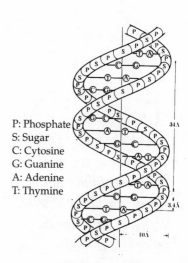

P: Phosphate
S: Sugar
C: Cytosine
G: Guanine
A: Adenine
T: Thymine

DNA molecule

THE GEOMETRY OF CHESS

When the world's sixty-four transformative stages represented by the "hexagrams" of the *I Ching* were arranged as an 8 × 8 square you may have noticed their resemblance to a chessboard. The ancient game of chess is a cosmological model in the form of a board game in which the opposing forces of the universe are brought to a more personal level. The chessboard, the spider's web, the Breath of the Compassionate, and the *I Ching* each in its own way represents

Thirteenth-century Spanish painting of a Christian and a Muslim playing chess.

The chess board is the world, the pieces are the phenomena of the universe, the rules of the game are what we call the laws of Nature. The player on the other side is hidden from us. We know that his play is fair, just, and patient. But also we know, to our cost, that he never overlooks a mistake, or makes the smallest allowance for ignorance.

—Thomas Henry Huxley

the world's opposite forces weaving the eightfold "elements" and qualities. The board's sixty-four alternating black-white (or red-white) squares represent the world's polarities, the warp and weft of matter's web, the compassionate cosmic exhale-inhale, the *yang-yin* interplay in sixty-four possible stages of transformation, the conflicting choices that life presents us.

The game of chess was consciously designed to represent our world of transformations on a restricted field of action. The board represents a Monad, the universe, a planet, a community, a person, a whole on any level. The game's pieces represent its elements and their possible actions in the world. The rules of chess are the rules of life, including those governing our own inner nature. When the Caliph of Baghdad was asked, "What is chess?" he answered, "What is life?" Indeed, chess has been called the "royal game of life." Chess represents the macrocosm and microcosm, the interactive opposing forces and principles of the universe and their correspondence to humankind's inner life and all its conflicting forces and possibilities. Understood in its original sense the game provides us with a way to discuss our own transformations.

It is believed that the game of chess was created around the sixth century either in India or Persia but, as pieces dating from the second century have been found in Asia, its origin may be much older. The 8×8 geometry of the board corresponds with the foundation geometry of a Hindu temple dedicated to the divinity in the form of Shiva, the transformer. The interplay among chess pieces represents the cosmos and our lives as fields of transformation.

In Zoroastrian Persia chess played out the struggle for the world between good and evil, light and dark, understanding and ignorance, as symbolized by the god or principle Ahura Mazda and his shadow twin Ahriman. Some Persian terms have been retained, such as "checkmate," which derived from *shah mat,* signifying "the king is dead."

The chess pieces represent forces of nature around us and psychological motivations within us; relationships among pieces on the board represent aspects of ourselves. Thus, chess gave the ancient philosophers a way to speak about inner development in symbols and to explore various possible psychological situations represented by the game's structure and process. Over the centuries, some pieces, including elephants and chariots, were different from the ones we know today. When chess arrived in Europe after the eleventh century, the pieces took on political and religious attributes.

Each step on the board, represented by the movement of the pieces, takes us through the polarities of the Dyad, light and shadow, the eternal struggle between consciousness and unconsciousness, wisdom and ignorance, our constructive and destructive motivations. We walk among opposite forces waging battle for domination of the world and Self,

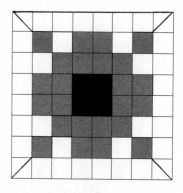

Hindu temple foundation geometry for a temple dedicated to Shiva, "the transformer."

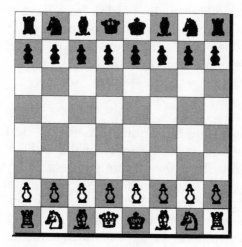

engaging in the struggle for spiritual enlightenment that is the unseen significance of each of our lives.

While each side begins with sixteen (= 2 x 8) pieces, there are only six types: pawn, rook, knight, bishop, queen, and king. They are another form of the six lines of the *I Ching*'s hexagrams, representing position, power, and possibility. Each piece begins the game on a specific location that determines its initial relationships.

Possibilities emerge rapidly: there are 20 possible opening moves, 400 second-move possibilities, and 20,000 possible third moves branching from their combinations. The pieces move around the board according to fixed rules, some remaining on one color throughout the game, others alternating colors beneath them.

A key to understanding the deeper significance of each piece is the *geometric way* it moves, whether in a straight line, as if along the side of a square, or diagonally along the hypotenuse of a triangle, or both. A "square" move represents earthly action. A triangular move represents alignment with the divine within us.

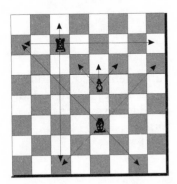

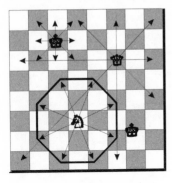

Each chess piece moves geometrically. The bishop, rook, and pawn travel along the edges or diagonals of a square. The king, queen, and knight extend their different powers through the world board by moving in any of eight directions, forming three different kinds of octagon.

The king, the most important piece, represents the sun of this solar system, our deeper or Higher Self, the divine spark within us. Yet it is the least powerful piece in terms of movement on the world board, traveling only one step at a time along both square and triangular lines in any of eight directions. The king is virtually hidden from the action, yet the entire game revolves around it. The king cannot be captured but loses when it is surrounded and is therefore totally restricted from movement on the board.

The queen is the most powerful piece on the board, having unlimited movement in any of the eight directions of manifestation. She is the mother goddess, the queen of the world, the genetrix in her natural "square" realm, the materially configured world. In another sense she is Regina Coeli, Queen of the Heavens, the widely traveling moon which always reflects the light of the sun, the king.

Each of two bishops, or elephants, remains on its start-ing color throughout the game and has the power to slide along divine diagonals through the world. In one interpre-tation, the white squares represent the path of intellect (called *jnanayoga* in India) while the black (or red) squares represent a devotional path of service through the heart (*bhaktiyoga*).

The two knights, representing the awakening spiritual initiate acting in the world, move by leaps of intuition along the sides of the thirty-sixty-ninety right triangle, the one found within a hexagon and which Plato called the most beautiful of triangles. The knight moves by alternating between white and black squares, head and heart, intellect and devotion. At most it can choose from among eight moves, eight directions in the world forming the corners of an octagon. It moves through the foursquare world by its own efforts, the only move the queen, the World Mother, cannot make for it. Only a knight or pawn can initiate the game by making a first move. Until then the world waits in quiescence.

🨂

The two rooks (castles or chariots) are the only pieces permitted to move strictly along straight lines on the side of a square in the four cardinal directions, representing our physical power to act in concert with the world's material structures.

♟

The eight pawns represent ordinary men and women attempting to cross the board of life through seven grades of initiation, mirroring the purification of the seven *chakras*, or centers of motivation. The Pawn can be male or female, rep-resented by the Roman Mercury and Venus or the Greek Hermes and Aphrodite, who merge as the pawn Hermaphrodite. A pawn experiences only the simplest interactions with other pieces and does not see the divine forces behind it, being aware of only one step ahead (with an option of two steps on its first move). It is the only piece that can move only forward, never backward, as all other pieces can. When presented with a challenge, the pawn has the possibility of worldly or divine action, probing straight ahead to the opposite color or along a short diagonal of the same color to capture or conquer its shadow opponent in an adjacent column. When a pawn has triumphed over the world's ordeals and reaches its seventh stage, the board's eighth square, the opponent's divine back row, it achieves a higher state, or "octave." Paradise is then regained and the pawn transforms into any piece the player wishes. Usually the pawn is made a queen and thereby becomes a co-creator with her, although great chess matches have been won by choosing to transform the pawn into an initiate knight.

A complete game of chess is an evolution through a series of geometric transformations of position and power. Every board situation is a snapshot from an ongoing process representing the characteristics of the moment, like the hexagrams of the *I Ching*. Each combination of pieces in various positions on the board exhibits the physical configuration of a particular energy pattern. Use your imagination to visualize their relationships as lines of force composing an energy web. One way to play the game is to work with the geometric tensions and make moves that bring the situation into harmony.

In the cosmology of chess, the four-sided board symbolizes the "world mountain," so we can look at it as four concentric rings of squares. The game's battle is often for the board's central area, the "high ground," the area of the Hindu temple known as the "seat of Brahma," the central chamber housing the divinity.

The board is ringed by twenty-eight squares and is therefore traditionally associated with the twenty-eight "mansions," or nightly stations, of the moon's orbit around Earth, recognized in many cultures. The number twenty-eight is interesting in that it results from the sum of numbers one through seven (28 =1+2+3+4+5+6+7) and was called a "perfect number" since, like the number six, it is one of the

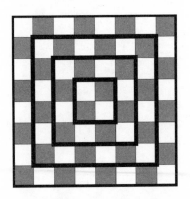

The chessboard as the earth-square, with the twenty-eight stations of the moon as its perimeter.

few numbers known to be equal to the sum of its divisors (28 = 1+2+4+7+14). It was also known as the number of days in a traditional lunar month of four seven-day weeks (28 = 4 × 7), thirteen of which make a lunar year of 364 (= 13 × 28) days, the time the moon takes to pass through the cycle of the zodiac. Thus the chessboard is circled by the moon in the same way that the moon's orbit encloses the squared circle containing the Earth (see chapter six). This is the limit of our action, the sublunar world, the "ring-pass-not" beyond which we cannot move our pieces.

The same twenty-eight-part symbolism has occurred throughout the ancient world going back at least as far as the Egyptian civilization. In terms of measure, the Egyptian cubit equals the width of seven "palms," or twenty-eight fingers. This symbolism extended to a colossal statue of Osiris, the Egyptian Lord of Cycles and of the Moon. Found near the Sphinx, the statue was made of twenty-eight pieces, the moon's monthly journey. Later, in Britain, twenty-eight boroughs were arranged to surround the square mile of central London as part of the geomantic system mirroring the cosmological patterns on earth.

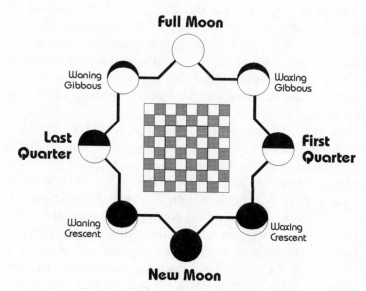

The phases of the moon

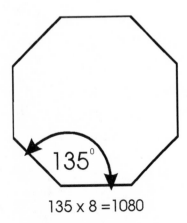

135 x 8 = 1080

The eight 135° corner angles of an octagon add to 1,080, the radius of the moon in miles.

Since the moon is a symbol of reflected light, the game of chess symbolically takes place within the orbit of the moon, as images reflecting archetypal patterns. The Octad and eight-sided geometry has often been associated with the moon, with water, and with lunar deities.

Throughout the world we find the octagon symbolizing the eight major phases of the moon's cycles showing the mirror symmetry of light and dark like the *yin-yang* of the *I Ching* in eight trigrams, or stages of transformation. Although the connection between the moon and the number eight is very subtle, it indicates the depth to which the ancient mathematical philosophers probed into the archetypal patterns and found them in nature. If we look at a regular octagon we find that each corner angle measures 135 degrees. The sum of all its corner angles must be 8 × 135 or 1,080 degrees. This number is the radius of the moon in miles, the sacred measure used to coordinate the cosmic dimensions and rhythms with the structure-function-order of terrestrial society. The moon's association with the tides and water accounts for the use of the octagonal shape in Christian baptistries, fountains, and fonts used in the ritual of anointing with water. John the Baptist appeared in ancient Babylon as Ioannes the Fish, who baptized initiates in the waters of their own unconscious, the lunar world of reflected images and archetypal patterns, in preparation for baptism by "fire" with Shamash, the sun god, and direct vision of the spiritual light "behind" the archetypes. The game of chess, a much later development, represents the journey within our own psyche and the battle to purify the unconscious.

Just as gold is the metal whose color, and inability to tarnish, has traditionally associated it with the sun, silver is linked with the moon. Indeed, on a clear night in the country the full moon appears quite silvery. The ancients chose a symbol more appropriate than they may have known. It was only in this century that scientists probing the atom found that the atomic weight of silver, its weight relative to carbon, is 107.870, or nearly 108, a tenth of the moon's radius of 1,080 miles. These numbers keep recurring not because we make them do so but because they are inherent in the proportions of nature that express the timeless mathematical archetypes.

Octagonal drinking fountain

LUNAR GEOMETRY IN ART

While some cultures see a "man in the moon," more universal is the image of a "hare in the moon." It was the hare-headed Teutonic moon goddess Oestra who gave her name to Easter, the celebration of resurrection and renewal that mirrors the moon being reborn by shedding its shadow.

This eight-sided bronze mirror back from the Tung dynasty depicts the "hare in the moon" using a mortar and pestle—symbols of male and female—to prepare the legendary "elixir of immortality." Its octagonal geometry implies the moon and renewal.

Place tracing paper over the picture and connect the corners and crossings of the interlaced squares to see the guidelines its designer used to achieve its internal proportions and placements.

As we climb toward the Decad we see how the principles of lower numbers serve as foundations. The principles of eight, the Octad, are a combination woven from the principles of its divisors, one, two, and four, the Monad, Dyad, Tetrad. The Octad displays wholeness, cycles, polarity, and manifestation. It serves as the principle of self-renewal at a

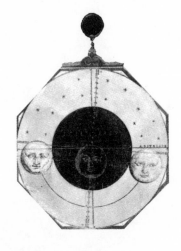

Representation of a lunar eclipse within an octagon (Apianus, *Astronomicum Caesareum,* 1540)

Tung dynasty mirror back

Aztec calendar stone

higher stage. The study of eight, therefore, gives insight into the mystery of the "same but different."

The Octad's role in the manifestation of the world is intertwined with the principles represented by all numbers, not just its divisors. With the introduction of the Octad into the universe that we're symbolically constructing, we gain the ability to apply self-renewal and limitless growth. We're limited only by the ever-expanding horizon ahead of us. But what is the nature of this horizon? Where *must* the Octad lead us?

Chapter Nine

Ennead

The Horizon

Playing under the earth, nine winters long, we grew mightily.

> —*Eddic poem*

Nine worthies were they called.

> —*John Dryden*

And when Abram was ninety and nine years old, the Lord appeared to Abram, and said unto him, I am the Almighty God; walk before me, and be thou perfect.

> —*Genesis 17:1*

And about the ninth hour Jesus cried with a loud voice, saying, ELI, ELI, LAMA SABACHTHANI? that is to say, My God, my God, why has Thou forsaken me?

> —*Matthew 27:46*

Facing a wall for nine years.
Neither being nor nonbeing stand; the universe is ultimately
  empty.
Nine years facing a wall, who is there that knows?

> —*Taoist poem*

Possession is nine points of the law.

> —*Common saying*

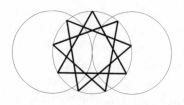

The Birth of the Nonagon.

THE WHOLE NINE YARDS

We've traversed the mathematical landscape and have come to the number nine, the greatest single-digit number within the Decad. The last among the seven Pythagorean numbers (three through nine), nine is the limit to which the generative principles of number reach. The ancient mathematical philosophers called nine the "finishing post" and "that which brings completion." It is the final step before ten, the Decad, a form of the Monad and beyond the realm of number itself.

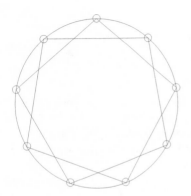

Composed of three trinities ($9 = 3 \times 3$), the number nine represents the principles of the sacred Triad taken to their utmost expression. Nine was considered thrice sacred and most holy, representing perfection, balance, and order, the supreme superlative. In ancient Asia, gifts given in groups of nine were considered most respectful. In fact, the Chinese words for "gift" and "nine" are identical.

Nine is the final number having a specific identity. It represents the highest attainment to be achieved in any endeavor. Nine is the unsurpassable limit, the utmost bound, the ultimate extension to which the archetypal principles of number can reach and manifest themselves in the world. The ancient Greeks called nine "the horizon," as it lies at the edge of the shore before the boundless ocean of numbers that repeat in endless cycles the principles of the first nine digits. Nothing lies beyond the principles of nine, which the Greeks called the Ennead.

Folk sayings have carried the essential nature of nine over many centuries. We say "a cat has nine lives," after which its luck runs out. Tales of European witchcraft recount how a cat may become a witch at the age of nine. A "cat-o'-nine-tails" provided symbolic maximum punishment to its victims. To "go the whole nine yards" is to go all the way in an endeavor. We say we're "on cloud nine" to express our highest joy. To be "dressed up to the nines" is to be dressed completely and perfectly, omitting no detail. "A stitch in time saves nine" is a pithy way of saying that taking care of small problems as they arise enables us to avoid having to repeat the complete job (nine stitches) later. The ninth level is the proverbial nth degree.

But in the way of a bargain, mark you me, I'll cavil on the ninth part of a hair.

—William Shakespeare

The number ninety-nine is also used to express the horizon of the Ennead. Landlords sometimes give ninety-nine-year leases and occasionally offer a symbolic 999-year lease to tenants. A traditional ninety-nine-year prison sentence is sometimes imposed to symbolically convey an ultimate term. Having completed his earlier way of life with idols, Abraham was ninety-nine years old when the Lord spoke to him. Islam acknowledges the ninety-nine Beautiful Names of God, the limit to which we can know Deity, represented by the ninety-nine beads of an Islamic rosary. The final word of prayer, "amen," from the Hebrew "so be it," transforms in the Greek numerical alphabet, the language in which the New Testament was written, into the number ninety-nine, signifying the highest extent to which human prayer can ascend before the infinite mystery of Deity.

Nine represents the boundary between the mundane and the transcendental infinite.

When mythological gods, goddesses, saviors and heroes went to extremes their actions were characterized by nineness. Nearly every tradition, from Northern Europe and Africa to shamanic Siberia, Asia, and the Americas, uses the number nine to express an ultimate extent, journey, or duration. Northern sagas are replete with ninefold symbolism. The Scandinavian god Odin, ruler of the nine Norse worlds, hung nine days on the world axis, or Yggdrasil tree, to win the secrets of wisdom for mankind. Every nine years, fertility feasts lasting nine days were held. Nine Norse giantesses, who strode nine paces at a time, lived at the edge of the sea and land. Prehistoric builders of the north recognized the symbolism of the number nine and built it into the structures of some stone circles.

According to Homer, the city of Troy was besieged for nine years. Odysseus wandered for nine more years after the battles at Troy before he finally returned home. The Greek goddess of the fruits of the earth, Demeter, who was often depicted with nine ears of wheat, searched nine days for her daughter Persephone before locating her in the underworld. As a result of her experience, Persephone is allowed to spend nine months of the year above ground when the earth is fruitful and for three months is compelled to live below when the earth is barren. The Greeks considered the under-

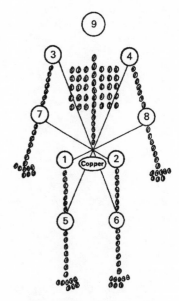

Diagram of the Dogon society in Mali as a skeleton made of feminine cowrie shells—Dogon money. The eight clans (joints) intermarry so that the two numbers always add to nine, the number of the head, the chief.

He rejoiceth more of that sheep, than of the ninety and nine which went not astray.

—Matthew 18:13

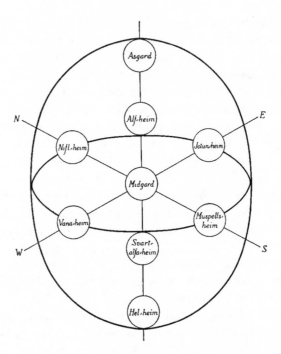

The nine worlds of the Odinic Mysteries.

Botticelli's drawing of Beatrice and Dante in Paradise viewing the circles of Heaven closest to God, numbering nine.

world to be so deep that, it is said, an anvil dropped into it would fall for nine days before reaching bottom. The ancient world's best-known mystery teachings, the Eleusinian Mysteries, were structured by the events of this myth: for nine days and nights the initiates took part in structured ritual and drama duplicating Demeter's search. The Greeks also told that the birth of Apollo and Artemis by Leto took nine days and nights and finally came about only when Hera sent her a gold and amber necklace nine meters long. The virtuous virgin Artemis, role model for young Greek women, chose two nine-year-old girls as her companions.

The Greeks honored nine *muses,* who personified and inspired the full range of the arts and sciences of humankind. As if to marvel at the results of their efforts, we accumulate them in *museums.* To be *amused* once meant to be under the sway of the muses. The Ennead was particu-

The Egyptian Ennead, or company of nine gods and goddesses, represents archetypal principles that regulate and rule the cosmos through the laws of number.

larly identified with the muse of dance, Terpsichore (from the Greek *terpsis* signifying "twirler," which later became "delighter in dance"), who begins at the Monad and dances through the nine digits, spinning around at the limit of nine, and dances back in this cycle.

Going back further in time, the Egyptians honored a company of nine "gods," or *neteru*, their Ennead of nine manifest principles, a trinity of trinities, which emerged from four pairs of unmanifest gods to oversee the world's ongoing creative process. The Egyptians saw the whole of humanity, which they symbolized by the glyphs for "nine archers' bows," as subject to the ninefold rule of the Ennead. The pharaoh, their representative on earth, was often symbolized by nine bows. This same symbol is referred to in the name of the vulture Mother Goddess Nekhebet, "she who binds the nine bows," or unites all people and so was called "the father of fathers, the mother of mothers."

Ninefold rule is not uncommon in history. In medieval Europe, nine types of crowns in heraldry represented complete ninefold rule. Ecclesiastical architects recognize nine styles of crosses. The Freemason founders of the United States government used the Egyptian Ennead as their model for the nine justices of the Supreme Court, whose interpretation of the law is ultimate.

The Egyptians honored ninefold companies of both deities and demons. The notion of ninefold cosmic levels spread through cultures they influenced and appears in others they never reached. In Christian symbolism we find nine orders of angelic choirs in nine circles of heaven and nine orders of devils within nine rings of hell. Medieval tales refer to nine gates of hell (three of brass, three of iron, and

The pharaoh came forth from between the thighs of the divine Nine.

—Egyptian myth

three of adamantine rock) with its nine rivers all pouring into the ninth ring, where Lucifer is confined. We are told by Milton that it took nine days for Lucifer and his angels to fall from heaven, expressing their ultimate distance from divine grace. The ninth represents the ultimate step. We learn that at the ninth hour of the clock Jesus died and that afterward death appears to his disciples nine times.

Nine was the central number of Celtic tradition, expressed as threefold aspects of the triple goddess, in myths of nine Celtic maidens and nine virgins attending Bridget. The sacred Beltane fire rites were attended by a cycle of nine groups of nine men. The culture is filled with references to nine.

Across the ocean the symbolism continues with the Native American, Aztec, and Mayan myth of nine cosmic levels (four above, earth in the center, and four below). In this context, nine-story temples also represent nine levels of the underworld, the realm of our unconscious.

The ancient Chinese told a myth of the man who emerged from the Yellow River carrying a geometric arrangement of forty-five points called the *lo-shu*, or river plan. The groupings of points were recognized as the numbers one through nine arranged in a 3 × 3 magic square. Its three columns, three rows, and two diagonals always add up to fifteen. Nine cells provide the smallest and first such arrangement in which all nine numbers can be woven in this form of matrix.

The ancient Chinese saw the magic square as a harmonious blend of the nine archetypal principles of number and a paradigm of cosmic structure and process, mirroring the supreme order of the universe. They used this arrangement to design the nine palaces of the Ming T'ang, or Hall of Light, in which the emperor lived and circulated every forty days, just as terrestrial energy circulates through the landscape in a year. In the science of *fêng-shui*, the numbers within each cell of the magic square have specific significance for working with the earth's subtle creative energies for the good of society and the environment. The divine central acre was dedicated to Shang-ti, the supreme cosmic ruler. The magic-square design was expanded and used as the layout of the capital city, with Old Peking at the center

Nine-story Mayan temple

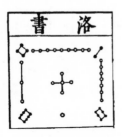

The 3 × 3 magic square appears in the cultures of Islam, Jains of India, Tibetan Buddhism, Celts, African, Shamanic, and Jewish mysticism.

having eight avenues of access. On an even larger scale it became the plan of the landscape of the whole of China, the Middle Kingdom of the world.

As the most auspicious number of celestial power in ancient Chinese, it orders nine great social laws, nine classes of officials, nine sacred rites, and nine-story pagodas. On the ninth hour of the ninth day of their ninth month was held the festival of the "double *yang*," honoring the ninefold creative celestial powers of heaven, the archetypes represented by the nine digits.

The 3×3 grid has been used elsewhere, notably in India, as the ground plan for temples dedicated to Vishnu, the Preserver. Each cell is subdivided into a smaller 3×3 grid, and each of the eighty-one $(= 9 \times 9)$ cells is dedicated to a different aspect of the deity. And we are told that a Buddhist monk's staff has nine rings.

In chapter three we encountered the worldwide practice of repeating a phrase, prayer, or chant three times to enforce it with sacred intention. Thus, to repeat any trinity three times, fulfilling nine repetitions, represents an ultimate appeal. A famous example concerns the three witches in *Macbeth* repeating their spell nine times.

A look into nearly every culture finds references to nine representing the ultimate extension. A superstition among many musical composers forbids the numbering of a symphony past the number nine. Even in modern times we find the principles of the Ennead expressed in secular affairs such as the American sport of baseball, played on the four-base diamond, or square earth-symbol, fielded by nine players over a duration of nine innings. The "bottom of the ninth," the modern version of the ninth rung of hell, is a team's last chance to win, the ultimate limit.

Beyond nine comes a new cycle. "Nine" and "new" are associated in various languages. The Egyptian glyph for "nine" was part of the glyphs for the sunrise and new moon. The word "new" in many languages derives from the Sanskrit word for nine, *nava*, later the Latin *nova*. The Roman market held every nine days was called the *novendinae*, and November was the ninth month of the Roman calendar (which became our eleventh month when January and February were added to realign the calendar with the proces-

Nine-level pagoda

*T*he weird sisters hand in
hand,
Posters of the sea and land,
Thus do go about, about,
Thrice to thine, and thrice
to mine,
And thrice again to make
up nine.

—William Shakespeare

*N*ow Peter and John
went up together into the
temple at the hour of
prayer, being the ninth
hour.

—Acts 3:1

Nine rings comprise the centriole of a human cell.

sion of the zodiac). A *novena* is an act of worship over a period of nine days. From the Latin *nonnes,* or prayers, at the ninth hour (which worshippers began counting at 3:00 A.M.) emerged the word "noon," the moment when the hands of the clock and the sun are in their utmost, or highest, positions.

There are few ninefold structures in nature. The forms that do configure the principles of the Ennead seem to be associated with the process of birth. The tail of the sperm half-cell is made of nine twisted threads. After it unites with the egg to form a complete cell, the first visible step in the doubling process of mitosis is the duplication of the centriole. Each centriole becomes a pole and moves to opposite sides of the cell while a spindle of chromosome threads forms between them. Electron microscope photographs show the structure of each centriole to be a circle of nine parallel tubes. The ancients could not, of course, have known about these discoveries of modern technology, but they attributed the nine months during which we develop in the womb and the nine openings of the body to the limiting principle of the Ennead.

THE HORIZON NUMBER

The source of the great esteem in which the ancients held the Ennead as the principle of completion, achievement, and the end of a cycle can be found in the arithmetic properties of the number nine. The best example is the familiar multiplication table, which we force children to memorize without an understanding of its inner structure. On its surface the multiplication table appears to be a random grouping of digits. But it is actually a map of the cross-fertilization of mathematical principles. To see how nine bounds it, we boil the table down to its digital roots. That is, we divide each term by nine and note only the remainder. Or we can simply add the digits together. The digital root of twelve is $1 + 2 = 3$. When the first sum is a two- (or greater) digit number, keep adding until you arrive at one digit. For example, $7 \times 8 = 56$ so fifty-six becomes $5 + 6 = 11$. Next we add $1 + 1 = 2$. The digital root of fifty-six is two. This process, called in

X	1	2	3	4	5	6	7	8	9
1	1	2	3	4	5	6	7	8	9
2	2	4	6	8	10	12	14	16	18
3	3	6	9	12	15	18	21	24	27
4	4	8	12	16	20	24	28	32	36
5	5	10	15	20	25	30	35	40	45
6	6	12	18	24	30	36	42	48	54
7	7	14	21	28	35	42	49	56	63
8	8	16	24	32	40	48	56	64	72
9	9	18	27	36	45	54	63	72	81

X	1	2	3	4	5	6	7	8	9
1	1	2	3	4	5	6	7	8	9
2	2	4	6	8	1	3	5	7	9
3	3	6	9	3	6	9	3	6	9
4	4	8	3	7	2	6	1	5	9
5	5	1	6	2	7	3	8	4	9
6	6	3	9	6	3	9	6	3	9
7	7	5	3	1	8	6	4	2	9
8	8	7	6	5	4	3	2	1	9
9	9	9	9	9	9	9	9	9	9

The familiar multiplication table and the multiplication table reduced to its digital roots by the process of "casting out nines."

Nine bounds decimals of infinitely repeating digits:

$$.11111\ldots \rightarrow 1/9$$
$$.22222\ldots \rightarrow 2/9$$
$$.33333\ldots \rightarrow 3/9 = 1/3$$
$$.44444\ldots \rightarrow 4/9$$
$$.55555\ldots \rightarrow 5/9$$
$$.66666\ldots \rightarrow 6/9 = 2/3$$
$$.77777\ldots \rightarrow 7/9$$
$$.88888\ldots \rightarrow 8/9$$
$$.99999\ldots \rightarrow 9/9 = 1$$

$$2 \times 9 = 18 \qquad 81 = 9 \times 9$$

$$3 \times 9 = 27 \qquad 72 = 9 \times 8$$

$$4 \times 9 = 36 \qquad 63 = 9 \times 7$$

$$5 \times 9 = 45 \qquad 54 = 9 \times 6$$

medieval times "casting out nines," reveals each number's essential nature and the principles of Monad through Ennead, which are at their core.

The table now reveals fascinating patterns. Each digit is highlighted as it recurs in the table to show its unique structure within the whole. Some columns and rows include all nine digits but in different sequences. In others, as in row six, only the same digits—three, six, nine—recur. Complementary rows, those that add up to nine—one and eight, two and seven, three and six, four and five—form patterns that are mirror images of each other but turned at right angles.

Only the pattern for the number nine has no reflection. A square of four nines appears at the table's center, and then a wall of solid nines forms a boundary along the table's edges, the proverbial horizon, or shepherd, which the numbers below approach and revolve before in patterns but never pass beyond. Nine bounds and directs the choreography of the cosmic order revolving within it.

Try expanding the multiplication table beyond nine and you'll see this same nine-cycle pattern of archetypes recur on larger scales.

The ancient Hindus developed the mathematics of nine digits and zero to regulate the rhythms of their religious

Multiplying by nine reveals a mirror symmetry among the numbers. When *any* number is multiplied by nine the resulting digits always add to nine. Because of this property the ancient Hebrews referred to nine as the symbol of immutable Truth, always producing itself and returning to itself, encompassing all numbers within its fold.

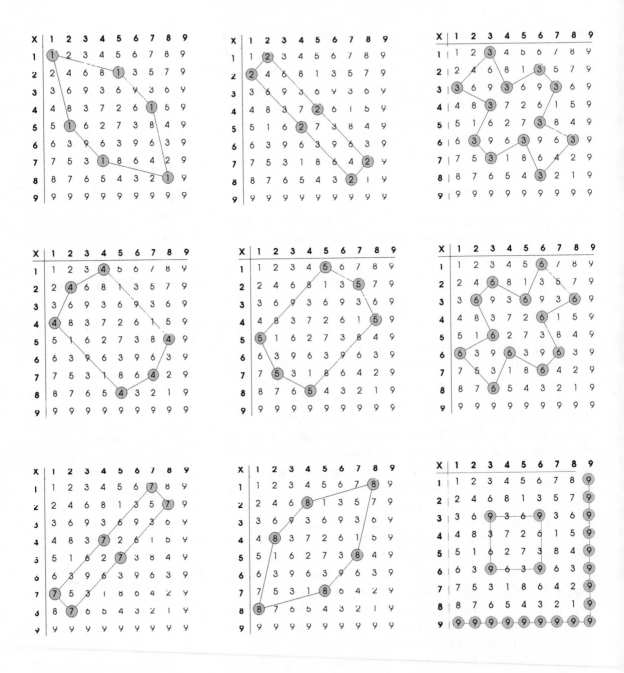

Each of the nine digital roots of the multiplication table forms a different pattern. Pairs of digits that total nine form complementary patterns turned at right angles to each other.

The unique digital-root pattern of the number nine bounds all the other numbers as their "horizon."

poetry, design temples, measure quantities and calculate astronomical positions to determine festival times. They used digital roots to check quickly the accuracy of large mathematical calculations, a process that was until recently taught in schools. For example, in the multiplication of 347 times 86, the first term reduces to $3 + 4 + 7 = 14$ and $1 + 4$ leaves a digital root of five. So does the second term, since $8 + 6 = 14$ and $1 + 4 = 5$. Their product, $5 \times 5 = 25$, reduces to a digital root of seven. Thus, they knew that the answer has a digital root of seven. And so it does: 29842 becomes $2 + 9 + 8 + 4 + 2 = 25$ and $2 + 5 = 7$.

$$
\begin{array}{rcl}
347 & \rightarrow & 5 \\
\times\ 86 & \rightarrow & 5 \\
\hline
29842 & \rightarrow & 7
\end{array}
$$

The Ennead bounds the regular polygons. If we look at the *sum of all the corner angles* within any polygon we find nine as its digital root. Nine serves to bound or enclose numbers and shapes despite their apparent differences.

As we look across the ocean of endless numbers toward an unseeable infinite, we really need not look past nine to find the horizon of mathematics.

The Ennead "flows around the other numbers within the Decad like an ocean."

—Nichomachus of Gerasa

X	1	2	3	4	5	6	7	8	9
1	1	2	3	4	5	6	7	8	9
2	2	4	6	8	1	3	5	7	9
3	3	6	9	3	6	9	3	6	9
4	4	8	3	7	2	6	1	5	9
5	5	1	6	2	7	3	8	4	9
6	6	3	9	6	3	9	6	3	9
7	7	5	3	1	8	6	4	2	9
8	8	7	6	5	4	3	2	1	9
9	9	9	9	9	9	9	9	9	9

X	1	2	3	4	5	6	7	8	9
1	1	2	3	4	5	6	7	8	9
2	2	4	6	8	1	3	5	7	9
3	3	6	9	3	6	9	3	6	9
4	4	8	3	7	2	6	1	5	9
5	5	1	6	2	7	3	8	4	9
6	6	3	9	6	3	9	6	3	9
7	7	5	3	1	8	6	4	2	9
8	8	7	6	5	4	3	2	1	9
9	9	9	9	9	9	9	9	9	9

X	1	2	3	4	5	6	7	8	9
1	1	2	3	4	5	6	7	8	9
2	2	4	6	8	1	3	5	7	9
3	3	6	9	3	6	9	3	6	9
4	4	8	3	7	2	6	1	5	9
5	5	1	6	2	7	3	8	4	9
6	6	3	9	6	3	9	6	3	9
7	7	5	3	1	8	6	4	2	9
8	8	7	6	5	4	3	2	1	9
9	9	9	9	9	9	9	9	9	9

X	1	2	3	4	5	6	7	8	9
1	1	2	3	4	5	6	7	8	9
2	2	4	6	8	1	3	5	7	9
3	3	6	9	3	6	9	3	6	9
4	4	8	3	7	2	6	1	5	9
5	5	1	6	2	7	3	8	4	9
6	6	3	9	6	3	9	6	3	9
7	7	5	3	1	8	6	4	2	9
8	8	7	6	5	4	3	2	1	9
9	9	9	9	9	9	9	9	9	9

The combined patterns of complementary pairs of numbers form unique arrangements of twelve (or twenty-four) points.

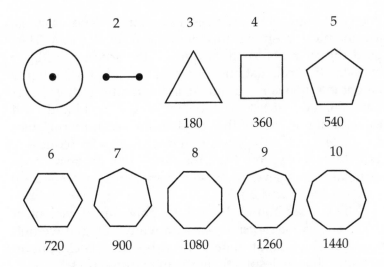

NINE POINTS

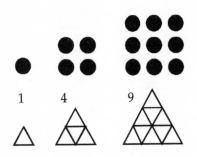

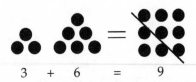

The ancient mathematical philosophers were keenly interested in the geometric shapes associated with different numbers. Into how many different symmetric shapes can you arrange nine coins? You'll find many ways, but four stand out: the square, cross, circle and triangle.

Arranged as a 3 × 3 square, nine points form a triple trinity, the highest expression of the sacred Triad. Just as the number three and its triangle are firstborn complete and whole among numbers and shapes, three times three as nine is the final child of Monad and Dyad, the ultimate completeness and wholeness. Along with one (ever-present unity) and four, nine coins configure the third and final *square* number we'll encounter within the Decad. Like all squares, nine is the product of one number (3 × 3) and sum of two consecutive "triangular" numbers (3 + 6 = 9). Nine, the ultimate number, happens to be unique as the *only* square number composed of two consecutive "cubes" (1 + 8 = 9).

The association of the square and cube with the number nine tells us that the principles of the Ennead are linked with those of the Tetrad, archetype of "earth," materialization, matter, configuration, form. Nine represents the ultimate expression of birth. We often encounter this association in the worldwide symbolism of triple goddesses, each with three aspects or articulations.

Four interesting ways to arrange nine points.

THE WORLD LABYRINTH

A second arrangement of nine points forms an X, or St. Andrew's cross, creating a different way of expressing the Ennead's bounding principles. From this nine-pointed core we derive the secret for constructing the mythic labyrinth found as symbol and architecture in Egypt, Sumeria, Greece, Crete, Rome, Oceania, Gothic cathedrals, and elsewhere. Labyrinths designed by similarly arranged five and thirteen points have been discovered.

A labyrinth is not a maze. A labyrinth has a single route that rhythmically spirals to the center. A maze offers multiple paths and is designed to confuse the traveler unless its secret is known.

Like the spider's web, a labyrinth represents the weave of the world. Our motions are bounded but guided by its walls. Sometimes built underground, a labyrinth symbolized the initiate's passage through the world, from profanity at its periphery to redemption at the sacred center. In this traditional labyrinth the traveler reverses direction nine times through eight rings, cycling in rhythms of left and right, outward and inward. Reaching the mystic center at the eighth ring, the pilgrim completes the octave and transforms to a higher, or deeper, realm, the core of Self.

Follow these steps on a separate piece of paper to create this labyrinth.

(1) Draw the cross of nine points. (2) Draw a vertical and horizontal line through the center. Draw two vertical and horizontal lines from each of four symmetric points as shown. The resulting figure is a cross. Draw a curve from the top of the center line as shown to the top of the line to its right. (3) Draw a curve up and around to the point in the upper right corner. (4) From the point at the upper left draw a curve up and around to the first line encountered at the

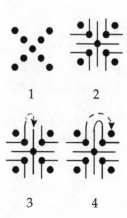

1   2

3   4

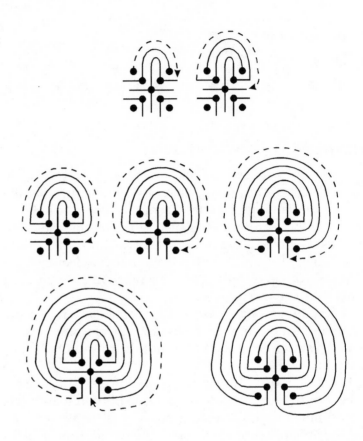

*P*oor intricated soul!
Riddling, perplexed,
labyrinthical soul!

—John Donne (1572–1631,
English poet)

left. (5–10) Follow this procedure of moving counterclockwise to each point or line and drawing a clockwise curve to the next point or line encountered at the right. The result is the traditional labyrinth design.

Initiates ritually traversing labyrinths chanted tones as they followed the path. The lost clues to their music are built into the labyrinth's structure. Each ring of this labyrinth corresponds to a note of the musical octave. The order in which the traveler ritually traverses the different rungs determines the sequence of notes to be chanted as the "song" of that labyrinth. Follow the path with your finger to find yourself being led through the sequence E-D-C-F-B-A-G-c[1]. If you play a musical instrument listen to this sequence to hear the labyrinth's musical theme.

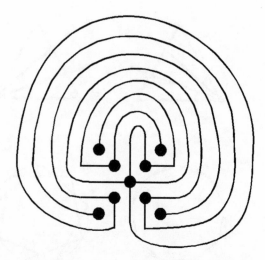

This traditional earth-labyrinth represents the initiate's journey through the "underworld" of the unconscious through a full musical octave to our spiritual core.

THE BIRTH OF THE NONAGON

Yet another configuration of nine equally spaced points, those arranged around a circle, gives us the corners of a nonagon, the nine-sided regular polygon. From the nonagon we can construct its "star" versions.

We can approach the horizon, but we never actually reach it. Like trying to trisect an angle, a precise construction of the nonagon is impossible to accomplish with only the geometer's three tools.

But because of the number nine's intimate relationship to three and six, any approach to a nonagon must be based on the geometry of the triangle and the hexagon. Follow these steps to manifest a very good approximation of the archetypal Ennead as nine points around a circle.

(1) Construct a hexagon within a circle (see chapter six), and inscribe its hexagram star. Choose one triangle as the nonagon's "seed." (2–4) From each of its three corners, draw

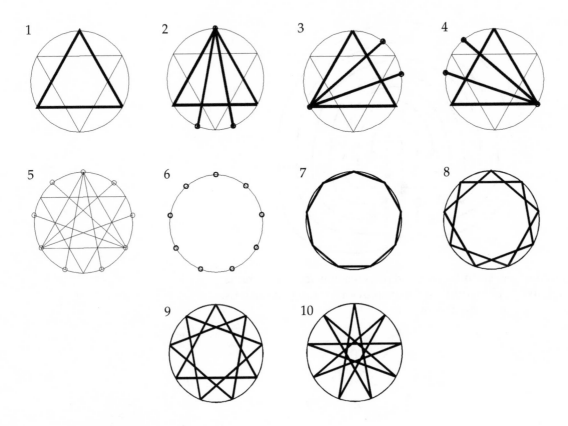

Islamic tiling pattern integrates the geometries and symbolism of the numbers three, five, nine, and twelve into one space-filling cosmological pattern.

two straight lines across to the opposite side through the points that cross the other triangle, extending them to cross the circle. Figure (5) shows all nine points and lines simultaneously. Figure (6) shows just the nine nearly equally spaced points. (7) Connect the points in sequence to construct the nonagon. (8) Connect every other point to form the dilated nonagram star, and (9) connect every third point to construct three noncontinuous triangles. (10) Connecting every fourth point produces the contracted form of the nonagram star.

THE ENNEAGRAM PROCESS

Another way to examine the principles of the ennead is through the arrangement of nine points around a circle. For this we look at the "enneagram," the name given to a remarkable diagram introduced to the West in this century by Georges I. Gurdjieff, who claimed to have learned it in an ancient monastery in Central Asia. It is remarkable in that it is a simple way of looking at whole events as an outer sequence with hidden inner connections.

The enneagram shows the organization of any whole event in terms of its essential principles. We've seen three outstanding archetypes associated with wholeness and cyclic process—the Monad, Triad, and Heptad, represented by the numbers one, three, and seven. They each contribute to any whole event. The enneagram weaves their principles into one diagram. We'll construct an enneagram as we apply it to a practical example, the functioning of a kitchen, based on the model by J. G. Bennett, a student of Gurdjieff's.

Begin by constructing a circle with nine equally spaced points. This is our Monad, a nine-spoked wheel representing the cycle of the whole process. In this example, a kitchen prepares and serves a meal, then cleans up and gets ready to repeat the procedure.

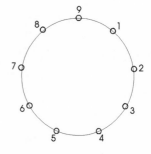

Put the numeral nine at the top, and number the points around clockwise. Draw straight lines connecting points three, six, and nine to form an equilateral triangle representing the principle of the Triad, the three-part structure

Neutral.
Circumstances,
conditions, place.
KITCHEN

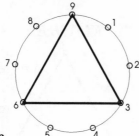

Active.
Something that brings
about transformation
or uses that which
is transformed.
DINERS

Passive.
Something that
undergoes
transformation.
RAW FOOD

inherent in every event—the passive, active, and neutral aspects. The passive element undergoes transformation. It is the raw food that will be prepared, cooked, and served. The active element brings about its transformation or uses that which is transformed. In this case the active element consists of the diners, the people who will eat the prepared food. The neutral element is the kitchen itself, which allows the polar opposites to interact in a constructive way, binding them into a complete whole at a new level.

These three aspects form the corners of the triangle with the raw food introduced at point three and the diners entering the scene at step six. The kitchen is at the apex at point nine. These three elements form the structure of any kitchen.

If we can identify the three structural aspects of any event—the active, passive, and neutral—we can begin to understand the way it works.

With these three guideposts we can fill in the intervening steps in the cycle. Each two consecutive steps extend the previous corner of the triangle. Steps one and two involve the kitchen. Steps four and five involve the raw food. And steps seven and eight involve the diners. Our experience in a kitchen enables us to fill in the stages to compose the outer sequence of events.

But the outer events also have nonobvious inner connections to other parts of the circle, to other stages in the process. We find them by applying the principles of the Heptad, the sevenfold harmony within whole cycles like the

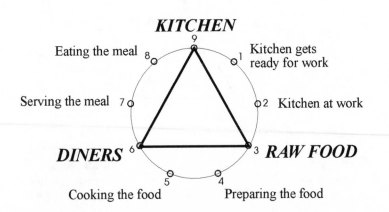

musical scale. The way that in seven seeks to return to unity is found simply by dividing seven into one (one-seventh). The result equals .142857142857142857..., a repeating decimal cycle of the six digits 1–4–2–8–5–7. This cycle contains all digits *except* multiples of three, which form the points of the Triad. Seven and three, Triad and Heptad, each supplies to the process what the other is missing. The Triad provides overall structure and balance but not the details of the process. The Heptad virgin guides the process but is devoid of the principle of balanced polarity. The enneagram binds them together, integrating all nine digits. (The circle of the enneagram itself represents the tenth digit, zero.)

Draw straight lines connecting the remaining points in this sequence (1–4–2–8–5–7), and draw a line from point seven to point one, renewing the cycle. These six lines are yet another way of expressing the principle of six circles around a central seventh, the six days of the week around the Sabbath, the six types of pieces on the chessboard and the permutations of the six hexagram lines of the *I Ching*.

These inner lines can help us to "think ahead" at any stage in the process. For example, let's say that we're at stage one, getting the kitchen ready for work. To determine what we'll actually do there, notice the line radiating ahead to point four. It tells us to look ahead to point four and consider what we'll need to prepare the food. So we gather cutting and cooking utensils, pans, spices, and whatever is needed to prepare the raw food. The line between points one and

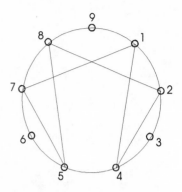

The fraction 1/7 = .142857142857... represented as a continuous path within a nine-pointed circle. Seven is the only number that divides into unity leaving a repeating trail of different digits.

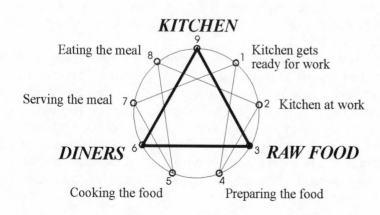

A man may be quite alone in the desert and he can trace the enneagram in the sand and in it read the eternal laws of the universe . . . If two men who have been in different schools meet, they will draw the enneagram and with its help they will be able at once to establish which of them knows more.

—P. D. Ouspensky (?–1947, Russian mathematician, writer, student of G. I. Gurdjieff)

seven tells us to consider what we'll need to serve the meal, so we also gather plates, eating utensils, napkins, drinking glasses, and so forth. Having thus prepared the kitchen, we proceed in the outer sequence to point two and get to work.

From point two we see lines radiating to points four and eight. Point eight has us consider the eating of the meal, the specific type of nourishment and any dietary preferences of these particular guests. It directs what we'll serve and how we'll serve it. Point four has us look at what we'll actually *do* to prepare the food, whether we bake, grill, fry, or steam it.

At point three the raw food is introduced, and we now can move to points four and five, where the food is prepared and cooked. Next, the diners are introduced, and at points seven and eight we can serve and eat the meal. Then, we clean up and reach the ninth stage, our goal of completing the cycle in preparation for repeating it later.

From each point, or stage, there are lines radiating to other points across the circle. These are guides to help us think out the process by directing us to look ahead and back to other stages. You can apply the structure of the enneagram to any cyclic whole event in your life if you can determine its inner Triad and the intervening stages that make the process a cycle. The Triad and Heptad are complements, each containing what the other is missing and comprehended in the nine-stage whole of the ennead.

Another example is the cycle of the year. But we must think of the yearly cycle as the ancients did, as a circle, relinquishing our image of the year as a straight line extending toward an incomprehensible "eternity." The ancients of many cultures represented the year as a circle, or spiral, whose dates recur with certain characteristics and resonate across time.

If we put January 1, the beginning of the Western year, at the top of the enneagram at point nine, the distance between each of the remaining points marks a period of forty days, a traditional milestone in time reckoning and mythological symbolism. The complete cycle comprehends 360 days, the traditional duration of the year. Consider the remaining five days as the Egyptians, Babylonians, Indians, and Mayans did, as intercalary days "outside" the circle of the year.

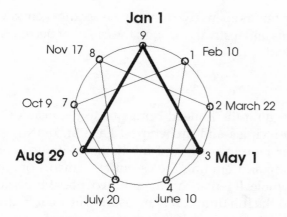

The inner lines of the enneagram show us how different parts of the year have inner connections. For example, perhaps you initiate an action on February 10. Taking action is like planting a seed that develops slowly, eventually producing flower and fruit in their seasons. Every process needs time to germinate and transform in its unique sequence. Its outer order around the circle based on the Triad shows us the event's familiar sequence in sidereal time. But the inner lines based on the Heptad show us how our actions ripple and weave across the circle of the year in *harmonic* time. Lines from point one, February 10, tell us that our action will have ramifications on June 10 and October 9. This system may sound far-fetched, but it works as an agricultural calendar whose Triadic structure of seed (January 1), flower (May 1), and fruit (August 29) indicate proper times for sowing, tending, and harvesting. The Heptad's inner lines show the necessary relations between them. In this structure the Egyptian calendar recognized the year as three seasons of four months each—Ahet, season of inundation and sowing; Pert, season of growth; and Shemu, season of harvest and inundation.

A look at the festival dates and religious holidays of many cultures reveals an emphasis on forty-day periods, grouped around each of the nine points of an enneagram. The Aztec and Mayan calendars, which contained eighteen months of twenty days, work nicely as enneagrams. Each of its nine points spans two months, so the inner cross-rippling of events and festivals through the calendar becomes easy to work with.

○

Reaching the Ennead brings us to the peak of the first nine principles and archetypes. Each encounter with the number nine in math and myth shows it surrounding or enclosing and binding the essential elements of any event into a whole. It represents the zenith of possibilities for number and the full duration of measure. The final Pythagorean number, nine, as the fulfillment of the Ennead, completes the archetypes that manifest the forms and processes of nature. With nine we exhaust the world's powers. As if after nine months in a womb, we are now ready for birth into life beyond which we have known. What remains for us is to go beyond the horizon, beyond the realm reachable by number and measure.

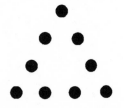

The clue to the way beyond numbers is found in the fourth configuration of nine points arranged as a triangle. The missing point implied at their middle will be our path through to ten, the Decad, the culmination of the cosmic process, not considered a number but existing as an empyrean realm beyond them.

CHAPTER TEN

DECAD

Beyond Number

Leaving the old, both worlds at once they view,
That stand upon the threshold of the new.

—*Edmund Waller*

The activity and the essence of the number must be measured by
the power contained in the notion of 10. For this (power) is great,
all-embracing, all-accomplishing, and is the fundament and
guide of the divine and heavenly life as well as human life.

—*Philolaus*

I will sing a new song unto thee, O God: upon a psaltery and an
instrument of ten strings will I sing praises unto thee.

—*Psalms 144:3*

My strength is as the strength of ten,
Because my heart is pure.

—*Alfred, Lord Tennyson (1809–1892, English poet)*

I could not tell nor name the multitude, not even if I had ten
tongues, ten mouths, not if I had a voice unwearying and a heart
of bronze were in me.

—*Homer*

And he wrote upon the tables the words of the covenant, the ten
commandments.

—*Genesis 34:28*

There are ten Sefirot. Ten and not nine; ten, and not eleven. If one
acts and attempts to understand this Wisdom he shall become
wise.

—*Sefir Yetzirah (The Book of Creation, third-century Kabalist text)*

Gone. Gone. Gone beyond. Gone beyond beyond. Hail the
enlightening voyager!

—*Sanskrit mantram*

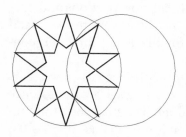

The Birth of the Decagon and
Decagram Star.

NEW BEGINNINGS

As we take the step to ten we pass beyond the ocean of numbers swirling within the ring of the horizon sharply defined by the number nine. But where are we when we pass the horizon? Ten takes us beyond the realm of number itself, above the fray of ordinary numerical interactions and geometric relationships. It is a new beginning, a journey into limitlessness.

Ten represents a recapitulation of the whole. It holds within itself the two parents of numbers (one and two) and their seven children (three through nine). Ten is a portrait of the whole family of archetypes gathered together, simultaneously displaying each of their principles. Expressing the properties of all numbers, ten represents a synergy, a whole greater than the sum of its parts, beyond the threshold of number itself. Like one and two, ten was not considered by the ancient mathematical philosophers to be a "number." Instead, it is the epitome of number, comprehending all numbers' principles as a mathematical promontory.

This all-encompassing archetype represented by ten has been known since the Golden Age of Greece as the Decad. The ancient mathematical philosophers were great punsters and played on the words *deka* ("ten") and *dachas* ("receptacle") to emphasize the Decad's property of containing the whole. They also referred to the Decad as "world" and "heaven," since it comprehends and harmonizes all numbers below it. Ten represents all-inclusive perfection. To understand the properties of ten is to know all. The Decad is a paradigm of the creating process, encapsulating all the archetypal principles of Monad through Ennead. It holds as if within ten fingers all their arithmetic proportions and geometric patterns that manifest as the cosmos. With the Decad we have all that is necessary for understanding the construction of the orderly universe.

To understand the Decad's principles we must look to its arithmetic origins. In chapter nine we saw how the essence of any number can be found through its digital root, the sum of its digits. The digital root of ten is one, that is, $10 = 1 + 0 = 1$. Thus, the Decad flows back to unity, becoming a new Monad. What the Monad is to the archetypal principles and numbers through nine, the Decad is to all numbers that fol-

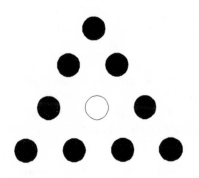

... ten ...
This number was of old
held high in honor,
for such is the number of
fingers by which we count,
... or the numerals
increase to ten, and from
there
We again begin a new
round.
—Ovid (43 B.C.–17 A.D.,
Roman poet)

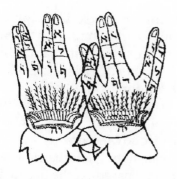 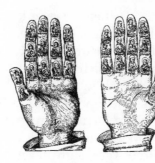

Two hands traditionally symbolize the cosmos in its entirety, summed up by the Decad of ten fingers. The left illustration shows the blessing by the hands from the mystical traditions of the Jewish Kabalah. Letters between the joints symbolize relationships between archetypal number/letter patterns, the human body, and the cosmic structure. The right hands show Christian saints in each of twenty-eight finger sections, a lunar reference.

low it. Multiplying any number by ten is like multiplying it by one. The result is essentially unchanged, but the number is brought to a higher level, an expanded version of itself. The Pythagoreans referred to ten as the "higher unity wherein the One is unfolded."

Another way of approaching ten is by looking at the origin of the word. The English word "ten" itself derives from the Indo-European *dekm*, signifying "two hands," a reference to our ten fingers, the most versatile parts of our body and the only ones capable of reaching any other part. From this came the Sanskrit *dasa*, the Greek *deka*, and the Latin *decem*. The Germanic branch changed the "d" to a "tz" sound (*"zehn"*) and from this came "ten." The clue of "two hands" shows us the factors that interact to generate ten. Since $10 = 1 \times 2 \times 5$, it results from the interplay of the Monad, Dyad, and Pentad, that is, the parents of number (one and two) themselves fertilized by the principle of regeneration (five). Thus, the Decad represents the power to generate numbers beyond itself, toward the infinite. Multiplying any number by ten does not change its essential nature but only acts to expand its power. Fifty is an expression of five.

Ten is the Pythagorean "perfect number," symbolizing both fulfillment and new beginnings. The Decad, ten and its multiples, brings all numbers to perfection and completeness. Thus, we praise something as "a perfect ten" and assign 10×10 or 100 percent as the highest grade on exams. Ten always represents the next generation, a quantum, or "logarithmic," leap ahead. When we want to express absolute comparisons and contrasts we say something is "ten times better" or "ten times worse" than something else and an army is "decimated" (literally "reduced to one-tenth"). When we refer to a *round* number we often imply a multiple of ten. Likewise, we group years in multiples of ten and name them *decades*, ten of which make a *century* and ten of those which compose one *millennium*. Celebrations like the Biblical "jubilee" were held in multiples of fifty (= 5×10) years. The Egyptian *sed* festival occurred every 30 (= 3×10) years of a pharaoh's reign to renew his spiritual powers.

Modern mathematics was greatly advanced when, in the Middle Ages, a greater need for calculating left the clumsy Roman numerals behind and became based on the decimal system of ten digits and place-value notation (of ones, tens, hundreds, and so forth), which is easy to manipulate. Yet decimal counting systems based on the ten fingers of our hands, the classic foundation of all counting methods, have always predominated around the world. Ancient cultures often had symbols representing large multiples of ten and methods of grouping them to express long durations of astronomical cycles and large quantities of weight, length, area, and volume. Even cultures without place-values (Babylonian, Egyptian, Roman), which used different symbols to represent numbers, grouped them by tens. The tenth, hundredth, and thousandth numeral are each unique, always taking the system to a greater level beyond the previous limit of nine (99, 999, and so on) with a new encompassing symbol. In modern form ten is the first value of two digits, and powers of ten (100, 1,000, and so on) extend another digit place, and these extend the archetype to new beginnings.

In counting systems worldwide, each tenth step begins a new level and recapitulates the whole. Number systems reveal a culture's picture of the cosmos.

Babylonian wedges are grouped in powers of ten and sixty.

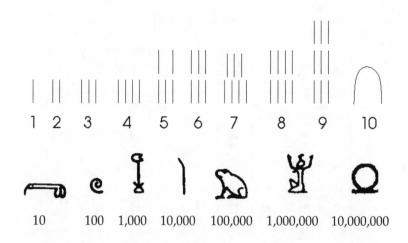

1 2 3 4 5 6 7 8 9 10

10 100 1,000 10,000 100,000 1,000,000 10,000,000

Egyptian line groupings led to the arch of ten, also represented by the glyph of the penis, indicating the Decad's generative power to create numbers beyond ten. Other glyphs (spiral, flowering plant, finger, frog [reference to myriad tadpoles], man in praise, and the eternal circle joined with a line) represented the greater fertilizing powers of ten.

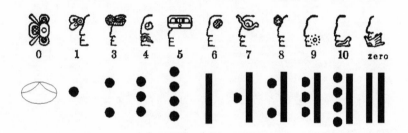

0 1 3 4 5 6 7 8 9 10 zero

Mayan dot and line represent one and five. The oval "zero" built powers of ten.

The familiar Roman numerals recur in powers of ten.
I II III IV V VI VII VIII IX X
L=50, C=100, D=500, M=1000

CHINESE-JAPANESE NUMERALS

1 一	5 五	9 九
2 二	6 六	10 十
3 三	7 七	100 百
4 ☐	8 八	1000 千

The Hebrew, Greek, and Arabic alphabets served as numerals arranged in a tenfold pattern of expansion. Each of these cultures encoded their great literature with the traditional number symbolism of ten archetypal patterns. This vertical arrangement of the Arabic alphabet also reveals the symbolic relationship between letters, numbers, and the four traditional elements.

Letters as numerals

א	ב	ג	ד	ה	ו	ז	ח	ט
Aleph	Bayt Vayt	Ghimel	Dallet	Hay	Vav or Waw	Zayn	Hhayt	Tayt
1	2	3	4	5	6	7	8	9
י	כ	ל	מ	נ	ס	ע	פ	צ
Yod	Kaf Khaf	Lamhed	Mem	Noun	Sammekh	Ayn	Pay Phay	Tsadde
10	20	30	40	50	60	70	80	90
ק	ר	ש	ת	ך	ם	ן	ף	ץ
Qof	Raysh	Seen Sheen	Tav	final Khaf	final Mem	final Noun	final Phay	final Tsadde
100	200	300	400	500	600	700	800	900

Hebrew

1-9	α	β	γ	δ	ε	ϛ	ζ	η	θ
10-90	ι	κ	λ	μ	ν	ξ	ο	π	ϙ
100-900	ρ	σ	τ	υ	φ	χ	ψ	ω	ϡ

Greek

Fire	ا	ه	ط	م	ف	ش س	ذ
	1	5	9	40	80	300	700
Air	ب	و	ى	ن	ض ص	ت	ظ ض
	2	6	10	50	90	400	800
Water	ج	ز	ك	س ص	ق	ث	غ ظ
	3	7	20	60	100	500	900
Earth	د	ح	ل	ع	ر	خ	ش غ
	4	8	30	70	200	600	1000

Arabic

THE PAST TENS

In ancient cultures number symbolism was consciously and consistently used to express alignment with the cosmic order. Throughout the world, myth and religion are replete with examples of the Decad as a symbol of fulfillment, expanded power, and new beginning. The appearance of ten often represents the completion of a journey and a return to the origin after a purifying ninefold experience. The Babylonians celebrated spring festivals at the time of renewal and rebirth by honoring their deities in a festival lasting ten days. Demeter searched nine days for her daughter Persephone, finding her on the tenth; the city of Troy was besieged nine years and fell in the tenth; afterward, Odysseus wandered nine years and returned home on the tenth. After nine days clinging to the cosmic tree, Odin gained wisdom for mankind beginning on the tenth day. After Dante toured the nine rings of hell and nine spheres of paradise, his experience culminated with an ascent to the tenth, or empyrean, realm of transcendent spiritual beauty.

In the Celtic traditions predating Christianity the burning of the sacred oak yule (from the Gallic *gule*, signifying "wheel") log took place on December 25. This ritual represented the annual wheel of the sun's path across the sky poised at the winter solstice between the old and new years. The yule log was accompanied by a stick of ash wrapped in nine bands symbolizing the past year's nine forty-day periods. The nine bands were lit first, burning and bursting one at a time. After the ninth band burst, the ash-wood was revealed to completely burn out the old year, which kindled the yule log soon to shine with the reborn light of the new year.

The Bible was purposely packed with consistent tenfold symbolism to indicate fulfillment, perfection, a leap to a new beginning that comprehends the past. The tenth generation after Adam was Noah, whose family became the world's new beginning of "one language and of one speech." Ten generations later came Abraham, whose faith was tested ten times and who, at age 100 (= 10×10), accepted God's covenant, gave birth to Isaac, and began the monotheistic Jewish religion, whose laws are encapsulated in the Ten

[T he ancients] used "ever-flowing Nature" as a metaphor for the Decad, since it is, as it were, the eternal and everlasting nature of all things and kinds of thing, and in accordance with it the things of the universe are completed and have a harmonious and most beautiful limit.

—Iamblichus

A nd her brother and her mother said, Let the damsel abide with us a few days, at the least ten; after that she may go.

—Genesis 24:55

W hen angry, count ten before you speak; if very angry, an hundred.

—Thomas Jefferson (1743-1826, third U.S. president)

Commandments. The Bible describes how Egypt was assailed by ten plagues before the Exodus began. Descriptions of the desert tabernacle during the forty ($= 4 \times 10$) years of wandering speak of ten-part structure and measure, as did Solomon's temple. The lyre with which David sang praises of God had ten strings to emphasize the transcendence of prayer.

In order for Jews to transcend individual prayer and to worship as a group, a quorum, or *minyan*, of ten men is required to hold a unified service.

Not only Jewish but Christian, Buddhist, Sufi, and Islamic traditions have Decalogues: there are ten Christian spiritual graces (love, joy, peace, long-suffering, gentleness, goodness, faith, prudence, meekness, temperance) and ten Sufi veils, or obstacles, that keep us from seeing the world's and our own divinity (desire, separation, hypocrisy, narcissism, illusions, avarice, greed, irresponsibility, laziness, negligence). The Pythagoreans recognized ten pairs of fundamental cosmic tensions that pull tight the weave of the universe (limited/unlimited, odd/even, one/many, right/left, male/female, rest/motion, straight/crooked, light/darkness, good/bad, square/oblong). Buddhists recognize ten perfections.

The ancient Egyptians, whose Pyramid Texts describe the Kingdom of Osiris (nature as perpetual cycles) upheld

Give thanks unto the Lord with the harp: with the ten-stringed psaltery do ye sing praises unto him.

—Psalms 33:2

The Endless Knot, one of the eight sacred emblems of Tibetan Buddhism, forms ten enclosures.

by ten pylons (the ten archetypes), described ten qualities characteristic of all created forms (activity, passivity, power, quality, quantity, relation, time, substance, position, peace). They divided the sky into thirty-six ten-degree arcs around the earth and honored the archetypes with a ten-day week.

Religions sometimes require a "tithing," or relinquishing, of 10 percent of one's income for the group.

The reader can find many more examples of tenfold symbolism in the religions and myths of virtually every culture. The consistency of this type of symbolism is not due to a cross-cultural sharing of information; it occurs to anyone who looks deeply enough into the properties of number.

THE TREE OF LIFE

One of the more well-known cosmological models based on the Decad is the Kabalah, the ancient system of mystical Judaism. Students of the Kabalah begin their study only after the age of forty (= 4 × 10), i.e., after the fulfillment of worldly concerns. It is an intricate system of letters, numbers, and sounds describing the structure of the cosmos, our body, and our inner life. The Kabalah approaches the twenty-two letters of the carefully constructed Hebrew alphabet and the mathematical archetypes from one to ten as divine instruments of the cosmic creating process. To those initiated in its wisdom it serves as instruction for conscious self-evolution.

The complexity of the Kabalah is also intended to keep it out of the hands of those not purified to a degree that will allow its proper use. In fact, the word Kabalah is at the root of the modern word "cabal," signifying a secret group. Although it was originally an oral teaching, some of the Kabalah's precepts were written in a third-century text called *Sefir Yetzirah*, or *Book of Creation*, which describes the unfoldment of the universe from the unknowable Ain Soph ("without limit"), the deity eternally beyond name, form, or description, the mysterious zero from which the creating process emerges. Always playing on words, students of the Kabalah point out that the word for "nothing" (*ain*) is composed of the same letters as the words "I Am" (*ani*). Thus declaring its Being, the Unknowable utters ten

*T*he Sacred Sefirot are Ten as are their numbers. They are the ten fingers of the hands, five corresponding with five. But in the middle they are knotted in Unity.

There are ten Sefirot. Ten and not nine; ten, and not eleven. If one acts and attempts to understand this Wisdom he shall become wise. Speculate, apply your intelligence, and use your imagination continually when considering them so that by such searching the Creator may be re-established upon his throne.

—Sefir Yetzirah

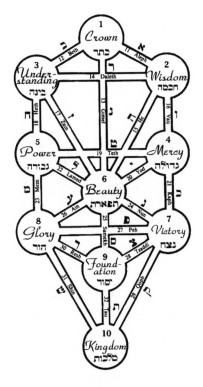

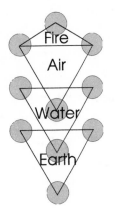

words, or archetypes, that generate the universe: Crown, Wisdom, Understanding, Mercy, Power, Beauty, Victory, Glory, Foundation, Kingdom. These ten "words," or Sefirot, are said to be the "potencies" by which the Divine manifests visible form. They are the ten archetypes of the cosmic creating process, described as ten lights or globes of luminous splendor that descend from the Divine arranged as three vertical columns ($10 = 3 + 4 + 3$). The names of these columns were Wisdom, Beauty, and Strength, after those of Solomon's temple, and were later taken as the motto of the Freemasons. Arranged in this manner the ten Sefirot are referred to as the Tree of Life, or Tree of Perfection.

The Kabalah teaches that we can only know the Unknowable God by his ten lights in the world, ten vari-colored emanations of divine qualities, like the visible flames from an unseen burning coal. In our experience of the world we cannot directly envisage the source of these lights but only their effects, the configurating manifestations of the ten archetypes, Monad through Decad. They are the vehicles through which the world's changes and transformations, the cosmic creating process, occur. Through deep study and contemplation of the ten numbers and an alphabet designed to correspond with the Divine unfoldment, students of the Kabalah gradually grow to know both the world and our deeper Self, finding them at one with their Divine source.

According to the teachings of the Kabalah, the first emergence from the Unknowable is the Monad, called the Crown. It further separates into a polarity, the Dyad, known as Wisdom and Understanding. These three form a divine triangle that gives birth to the seven archetypes or Sefirot, like the parents of number generating the seven numbers three through nine.

This divine upward-pointing triangle atop the Tree of Life manifests the remaining archetypes by reflecting itself as three inverted triangles, unfolding the world of ten spheres in four triangles, or levels, of increasing density, in other words, the four ancient elements, or modern states of matter, from electronic plasma to gas to liquid to solid (see chapter four).

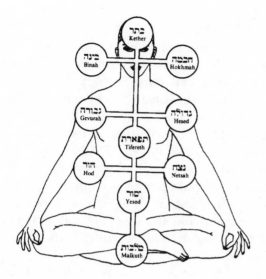

The ancient Jewish seers who developed this model show its interrelation with the alphabet to produce thirty-two connections, or paths, along this unfoldment. Thus, a divine descent along thirty-two paths to unfold creation plus thirty-two ascending paths returning to Divinity produce sixty-four stages of transformation, reminding us of the models of the *I Ching* and chessboard (see chapter eight).

The Tree of Life is sometimes overlaid onto the form of a human body, creating an image known as the Universal Man, or Adam Kadmon. The Tree of Life reveals itself to be a caduceus, the modern medical symbol with roots in the ancient past, depicting the three channels of our subtle nervous system (see chapter seven). Thus, the Decad and Tree of Life are seen by the Kabalah's students as being within us, a world axis on which the cosmos and our consciousness turn. In India, this symbol is known as the spine, or *shushumna*, flanked by the rising and falling "serpents" of creative energy, *ida* and *pingala*. Along this Tree of Life are the seven *chakras* distributed within the four levels.

The system of the Kabalah requires a lifetime of serious study and self-development with a qualified teacher. It is one of many descriptions of the cosmic creating process based on the principles of the ten archetypes of number.

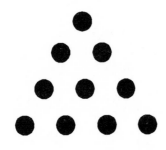

THE TETRAKTYS

Another cosmological model based on the completeness of ten points unfolding in four levels to describe the universal creating process was called the Tetraktys (from the Greek for "fourfold") in the Pythagorean school around 500 B.C. It consists of ten points or pebbles arranged as an equilateral triangle like bowling pins, with one, two, three, and four points in each row, since 10 = 1 + 2 + 3 + 4. The Tetraktys is a convenient and powerful tool for examining simultaneously all ten archetypal principles of number underlying the geometry that manifests itself as nature. It was held in the highest esteem, and, like the Tree of Life, it remained strictly in the secret oral tradition for centuries before anything was written about it.

The Tetraktys served as the basis of the Pythagorean school's studies of natural science and philosophy. Its triangular form indicated to the symbolic geometers its divine nature. They took its four levels to represent increasing densities of the four elements, fire, air, water, and earth, the four modern states of matter (electronic plasma, gas, liquid, and solid). This descent from "fire" to "earth" represented by one, two, three, and four points is a metaphor for the way the ten archetypes unfold and clothe themselves with matter to manifest as nature's visible patterns.

The Tetraktys also served as the core of their mathematical teaching. Its four levels show us the number of points in space of zero-, one-, two-, and three-dimensions, thus defining all possibilities of point, line, surface, and volume. The Pythagoreans saw the cosmos unfold in the same sequence as a plant grows: seed, stem, leaf, and fruit.

Fire ● Point
Air ● ● Line
Water ● ● ● Surface
Earth ● ● ● ● Volume

By studying the Tetraktys the ancient mathematicians noticed that ten is unique as the *only* triangular number among the infinitely many that is a sum of consecutive odd square numbers (1 + 9 = 10).

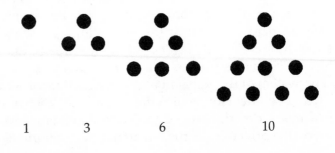

1   3   6   10

The Pythagoreans first explored the simple arithmetic and geometry in the Tetraktys. Using dark and light stones (you can use two different types of coins) to form patterns among the ten points, they saw that the regroupings revealed the relationships and patterns of interplay among simple numbers. But most important, they observed that the Tetraktys is a complete summation of three-dimensional forms because it holds the geometry of all five Platonic volumes, which divide space equally in all directions (chapter four). Every three-dimensional form is ultimately based on one or more of these five volumes. Through a triangle of ten simple points the Decad's principles of fulfillment, completion, and wholeness are revealed to the symbolic geometer.

It is also well known that the Pythagoreans were deeply interested in the mathematics of music, the simple ratios among whole numbers that produce the pleasant tones of the musical scale. The Tetraktys encapsulates the numbers one, two, three, and four as the fractional lengths of a vibrating string that produce the natural seven-tone musical scale: the octave (one-half and two-fourths) the double octave

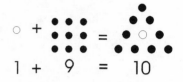

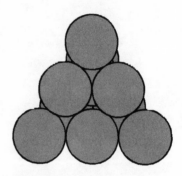

The Decad extrudes into three dimensions as a tetrahedron when ten spheres, like oranges or cannonballs, are stacked in rows of 1 + 3 + 6.

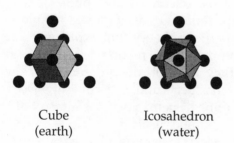

Cube
(earth)

Icosahedron
(water)

Octahedron
(air)

Tetrahedron
(fire)

Dodecahedron
(heaven)

(one-fourth), the musical fourth (three-fourths), and the musical fifth (two-thirds). These produce the "spiral of fifths" or "spiral of fourths" and endless octaves of tones (see chapter seven).

Members of the Pythagorean school were so impressed with the richness of this cosmological model that they swore oaths "by the discoverer of the Divine Unregenerate Tetraktys, the spring of all our wisdom, the eternal root of Nature's fount."

The Pythagoreans realized that the Tetraktys represented an ensemble, a unity, a summing up of the whole, that comprehended the completeness of mathematics and the archetypes that manifested themselves as the visible forms of the world. All ten archetypal number principles can be studied in it. For this reason mathematical philosophers have studied the Tetraktys for over twenty-five centuries to investigate the ten mathematical archetypes that construct and guide the natural course of events.

THE BIG T.O.E.

Mathematicians, scientists, and philosophers never stop looking for a unifying cosmology, a simple but complete scheme of nature, a theory that explains in one symbol or equation the universal creating process. Yet it seems that whenever they do look deeply enough, they come to a bedrock of ten aspects and rediscover the Decad regardless of their counting system, since it is a result of the interplay of the eternal number itself.

A recent wave sweeping theoretical physics has been called The Big T.O.E., or Theory of Everything. It proposes that the fundamental building "blocks" of the universe are neither "particles" nor waves but infinitesimally small "loops" of mathematical "string" that roll, rotate, twist, vibrate, and tweak to ultimately result in the familiar configurations of matter. Only the string's mode of vibration produces the universe's endless variety, determining all possible particles and forces of nature. The interesting point is that the design of these "strings" requires a mathematics that only makes sense in ten dimensions.

Simply put, string theory proposes that each of our familiar three dimensions of space (length, width, and height) is actually three dimensions enfolded to appear as one. Each dimension is treated mathematically as a Triad, a Borromean Ring tripartite unity. This Triad of Triads makes nine dimensions of space in all. The flow of time accounts for the tenth dimension. True to its ancient and archetypal symbolism, a Decad describes a whole.

While string theory reveals breathtakingly beautiful mathematical equations, there is no theoretical evidence for it. Its validity is presently impossible to detect and exists only in the uncannily consistent equations of its mathematics. While Pythagoras is credited with conducting the first investigations to apply mathematics to actual events by finding numerical ratios among musical strings, in string theory we are witnessing fifth-century B.C. mathematics finding rebirth at the end of the twentieth century, along with a recurring recognition of the archetypal principles found at the core of number and shape.

String theory postulates a ten-dimensional universe in which each of the three dimensions of space (length, width, and height) are a trinity, plus the dimension of time.

THE BIRTH OF THE DECAGON

The geometry of the Decad emerges from that of the Pentad and shares its arithmetic patterns and self-regenerating golden ratio. The fact that $10 = 1 \times 2 \times 5$ tells us that the decagon can be constructed as the doubling of a pentagon in a circle. While the ten-sided decagon is born through the *vesica piscis*, the Decad comprehends the geometries expressing all the archetypes from Monad to Ennead, including that of its own "parents." Follow these steps to construct the decagon and its internal patterns.

(1) First construct a pentagon and pentagram star (see chapter five). (2–3) From each of the five corners draw a straight line across the circle and through the point where lines intersect, extending each line to reach the circle. (4) The five corners of the pentagram plus five points indicated by the extended lines produce ten equally spaced points around the circle.

*E*ach thing is of like form
everlasting and comes
round again in its cycle.

—Marcus Aurelius

Connecting points around the circle in different patterns yields different configurations of the Decad. (5) Starting with the topmost point, connect consecutive points to construct the decagon. (6) Connect every other point to make two discontinuous pentagons; (7) connect every third point to create the "decagonal star" and (8) every fourth point to produce two discontinuous pentagram stars. You can use golden ratio calipers to explore its self-reproducing Φ ratio (1.618... to 1).

Use your imagination to create and explore its other internal geometric patterns by drawing straight lines between these ten points and points created at intersections. These patterns are the ones used by artists, craftspeople, and architects of all eras in designing works that express the principles of the Decad.

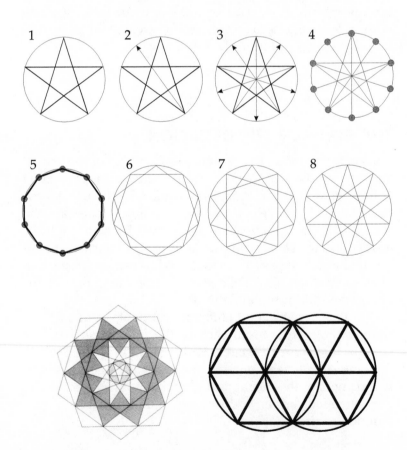

DISCOVER THE DECAD IN NATURAL FORMS

The principles of the tenfold Decad manifest themselves in living nature, often in organisms that display fivefold symmetries like microscopic diatoms and radiolaria. More highly developed creatures like lobsters and crayfish, cousins of the eight-legged spider, have ten walking legs. The squid, the animal with the most complex eye in the invertebrate world (sometimes mistaken for the eight-armed octopus), has ten tentacles.

Since the most widespread expression of fivefold principles is found in the plant world, we can expect to find tenfold geometry there, too. While some flowers, like the well-known passionflower (*Passiflora caerulea*), are considered to have ten petals, they actually have five petals subtended by five leafy bracts, giving the appearance of ten petals. But tenfold arithmetic and geometry can be seen in their fruit and seedpods, as in many roses and citrus fruits. Along with the Pentad's principle of regeneration, the Decad and tenfold symmetry are found where the plant holds its new beginnings and leaps to a greater level of fulfillment and completion of the whole.

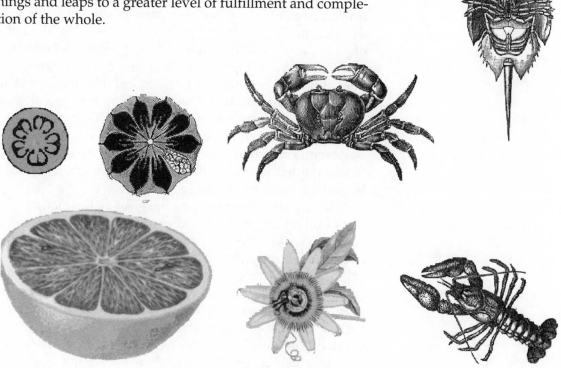

DISCOVER THE DECAGON IN ART

Construct an Islamic Tiling Pattern

We've seen elsewhere how the Islamic tradition uses geometry to express each of the universal archetypal patterns, and this use extends to the Decad. The following construction guides you to produce a decagon with an intriguing internal tiling pattern symbolizing the regeneration and fulfillment of enlightened humanity when it appears on walls, carpets, books, and elsewhere. All lines should be drawn lightly until the outline is completed and the significant lines can be darkened to emphasize the pattern.

○

(1) Construct ten points around a circle. Connect every fourth point to produce two pentagram stars. (2) Draw short lines within and parallel to the central decagram star to "thicken" it. (3) Place your compass point on one of the crossings (point a) of the pentagram stars and open it to the nearest corner of the inner decagram (point b). Lightly turn a circle, crossing the line at point c. (4) Now place the point of your compass at the original circle's center and open the pencil to point c. Lightly turn another circle, crossing the lines at point d and others like it around this circle. (5) With the straightedge draw a line connecting points c and d, extending it slightly beyond them. This will create a small regular pentagon pointing toward the inner decagon. Do the same with every pair of points around the circle. (6) Notice where the ten extended lines cross. Join them to create a large decagon, the outer edge of the tile.

This construction serves as the basis of the many variations of the tiling pattern and geometric ornamentation. If you wish to develop the tile in more detail, notice that only two patterns complete the Islamic design: a small four-sided "kite" containing a smaller replica of itself and a pentagram containing a five-pointed star surrounded by repeating irregular "T" shapes. With care it is possible to construct this pattern in the appropriate areas within the tile.

Reproduce multiple versions of this tile to see how they "fit" together by leaving "gaps" in interesting geometric symmetries.

TRÈS RICHES HEURES

Another of the many examples of the transcendent symbolism of the Decad is found in the illustration of Paradise in the famous *Très Riches Heures du Duc de Berry* in the fifteenth century. The scene depicts important stages in the story of Adam and Eve, from the serpent and the apple to expulsion from the garden. The composition is obviously designed within a circle, the traditional symbol of heaven, wholeness, and unity. The artist has actually left visible some of the underlying geometric frame with which the scene was composed. When a decagon, symbol of the Decad's transcen-

Paradise. Geometry after
Charles Bouleau (*The
Painter's Secret Geometry*).

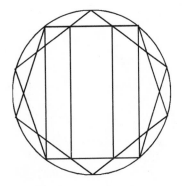

dence and fulfillment, is inscribed within the circle, the internal guidelines become apparent. Draw straight lines connecting every other point to construct two interleaving pentagons. Straight lines drawn in different patterns between the ten points reveal the scene's invisible grid. For example, vertical lines connecting the pentagons' corners and crossings define the horizon and the architecture of the fountain and also integrate the figures within the scene. Draw lines, or simply hold the straightedge between corners and crossings to discover the reasoning behind the placement of each element within the circle.

Finally, Adam and Eve find themselves expelled into a new beginning beyond the gate of the ten-pointed circle of Paradise. Yet even their placement in the world of toil emerges from the geometry of the decagon. The circle is actually inscribed within a square, implying that Paradise is imbedded and hidden within material manifestation. As always, extension lies along diagonals. Put your compass on the bottom left corner of the square and open the pencil point to its upper left corner. Turn a circle, or part of one, to see that even in their expulsion Adam and Eve are still embraced by the Paradise they left.

THE CATHEDRAL OF ST. JOHN THE DIVINE

We tend to think of cathedral building as an art of long ago, limited to a flurry of cathedrals built during a few centuries in the Middle Ages. But there still exists a small core of peripatetic stonemasons dedicated to preserving the European cathedrals. In fact, the world's largest cathedral is now in the process of being built entirely of stone on the highest hill on Manhattan Island in New York City. The cornerstone of the Cathedral of St. John the Divine was laid on St. John's Day, December 27, 1892, and the structure is expected to be completed near the end of the twenty-first century. While built primary of limestone, the cathedral also incorporates each of the twelve types of stone mentioned in the Revelation of St. John the Divine, the last chapter of the New Testament.

Proposed completion of the front (west facade) of the Cathedral of St. John the Divine.

Keeping the cathedral-building traditions alive, masons and sculptors come to New York City from around the world to pass on their ancient knowledge to young people who live in the community. But the stoneworkers' ancient techniques are now augmented by computers and laser-guided saws, although the detailed carving and placement are still done by hand.

From its inception to 1945, the cathedral was overseen by five architects, perhaps the most significant being Ralph Adams Cram of Boston, who was influenced by the French Gothic cathedral tradition, well known for the wonderful geometry of its architecture. While the architects left detailed blueprints for the cathedral's construction and the dimensions of every stone, they, like their predecessors, left no clue to any underlying geometric scheme they may have used to integrate and harmonize the whole design.

Like the Egyptian and Mayan pyramids, a cathedral is built from the ground up but is meant to be viewed as descending from heaven above, hanging down from the sky.

Cathedrals in general are among the greatest symbols of Divine manifestation in the world and of spiritual regeneration and return to the Divine within ourselves. The length of this cathedral is 601 feet, the length of an American football field plus one football; more important, the number 601 is associated with the Greek word *kosmos*, "cosmos," or "orderly universe." The geometric symbols of regenerative life and spiritual fulfillment are the pentagon and decagon, expressions of the archetypal principles of the Pentad and Decad. It is logical to look at the cathedral's underlying geometry as somehow related to these shapes. The following construction will help you to discover for yourself many of the geometric relationships integrating the elements of the front entrance, that is, the west facade of the Cathedral of St. John the Divine. The same approach will allow you to investigate the architecture of European cathedrals, although the seed-polygons will differ.

1. A rectangle has been constructed around the maximum width of the cathedral at its base and the full height to the tops of the towers. The crossing of this rectangle's diagonals show us the facade's center. A circle has been circumscribed around the rectangle.

2. Around this circle have been found the ten equally spaced points of a decagon, symbol of completion and transcendence. Every other point is connected to form two interlaced and opposite pentagons. Notice how the base of one pentagon matches the cathedral's base, while the other pentagon limits the height of the towers, whose innermost turrets touch the points made where the pentagons cross.

3. Slowly we can unfold the geometry of the cathedral. An upright pentagram has been drawn in one of the pentagons. Its crossings reveal the distance between the towers. The crucial placement of the rose window is found within a pentagram star inverted and lowered from the center of the large pentagram star.

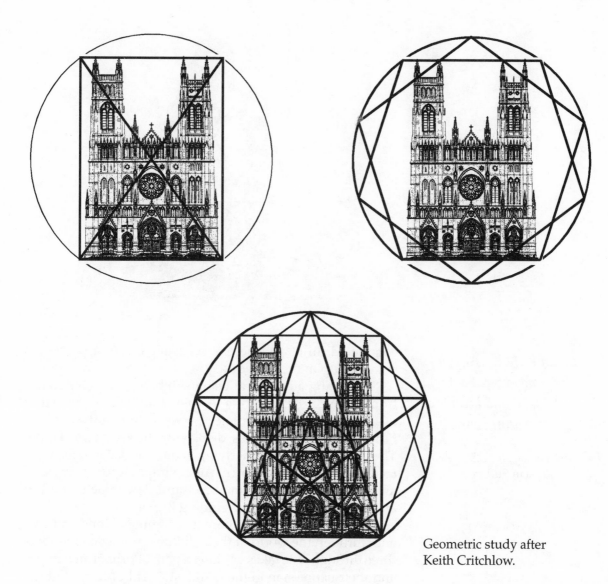

Geometric study after
Keith Critchlow.

The placement of other important elements and details
of the facade are left for the reader to find by completing
unseen lines guided by the points of the overall decagon.

Because the plan includes the geometry of the pentagon,
it may be "folded" into one half of a twelve-faced dodecahedron, the fifth Platonic volume and another symbol of
heavenly order. Its floor plan also does the same; together
they comprise a full dodecahedron, the heavenly sphere

*In an instant, rise from
time and space. Set the
world aside and become a
world within yourself.*

—Shabestari (c. 1250–1320,
Persian Sufi poet)

completed. In effect, the entire cosmic creating process is summed up in the cathedral.

Designing a cathedral using symbolic geometry comes from a timeless tradition of temple building and geometric design. The temple mediates between humanity and the heaven of archetype ideals, displaying the geometry of both. The discovery of an underlying geometric scheme in architecture, art, or nature can give us a glimpse into the world of archetypes. Within the imperfect and flawed actuality lies an inherent ideal invisibly guiding its way.

The journey from the parents of numbers one and two and, through the world of their children, the seven numbers three through nine takes us along a path of primal archetypes and their expression in the manifest world of form. But the step to ten is not so much a leap across a chasm as it is a recognition of the inherent unity and wholeness that have been present all along. The Decad seems to observe from beyond the fray of numbers, enclosing them and enabling them to extend beyond themselves to higher expression and fulfillment. The Decad is the recurrence of the Monad, unity at another level. With this new beginning we are back where we began, although the better for having gone through the experience of which the Decad is the summit.

Now That You've Constructed the Universe...

The geometer's aim, therefore, is to imitate the universe symbolically, depicting its central paradox by bringing together shapes of different geometric orders, uniting them as simply and accurately as possible and thus creating a cosmic image.

—John Michell

... what have you actually done? If you followed the steps for each of the constructions, you used the geometer's three tools (and your awareness) to represent the archetypal principles of the Monad through Decad, one through ten, the circle through decagon (plus the twelve-sided dodecagon and the five Platonic volumes), and you've acquired some practical skills. These are the constructions used by architects, carpenters, graphic designers, surveyors, and engineers. Remember to begin each construction with a central point while centering yourself within. Watch as each polygon emerges from the polar tension created by the two circles of a *vesica piscis*. With practice you should be able to reproduce each construction from memory. You'll see that certain numbers and their shapes are closely related, such as two, four, and eight; and three, six, nine, and twelve; and five and ten. Their constructions grow from related ones, and their principles are so closely allied that they teach us about each other.

The geometer contemplates these constructions and gradually becomes familiar with the principles they express by developing a feel for each geometric motion, the particular way it expands the design, and the unique relationships it produces among the lines, angles, areas, and volumes. In contemplating each construction the geometer becomes familiar with the message of each mathematical archetype, getting to know it as a different expression of harmony. With

347

enough experience the geometer can transfer this skill to recognize the character of relationships and bring harmony to other situations. The geometer learns to establish the calm eye within any storm.

Nature manifests herself in her organic and inorganic forms the principles of numbers as if through shaping filters or templates. The few characters of nature's alphabet recur in a consistent geometric design-language. All nature's shapes represent invisible forces, fundamental principles, and cosmic processes made visible. The geometer's step-by-step motions are a metaphor of the world's drama. The orderly process of geometric construction shows us the steps nature metaphorically uses to clothe the archetypal patterns. The geometer gradually learns to see the world of common experience as governed by laws of nature that can be understood as expressions of mathematical archetypes. Look at the forms you are most familiar with, like plants, a flower, pineapple, or pinecone, and practice seeing them not as "things" that simply popped into view but as the result of a gradual process of manifestation. Imagine each having begun as "invisible" light that shapes itself into geometric patterns of energy or lines of force that configure the guidelines upon which the visible matter precipitates. If you applied the geometric constructions to the illustrations of natural forms from leaf to flower and butterfly to starfish, you should soon be able to look at other natural forms and recognize their general geometric patterns. What characters in the geometric alphabet does that vegetable in the grocery store display? A triangle, square, pentagon, hexagon, spiral, or other shape? If your pencil was sharp and you further developed the constructions upon the illustrations, you may have been awed at the degree of detail to which nature follows her patterns. By learning to recognize nature's calligraphy and observing the world through its shaping principles, the geometer can read nature like a book and understand something of nature's intention. Deep within yourself you already know the cast of characters and the plot, the principles of number and shape that nature uses. The geometry serves to remind us of it. You probably know other examples of geometric nature not chosen here, and now you have the skills to explore them on your own. A

Nature will reveal itself if we will only look.

—Thomas Alva Edison
(1847–1941, American inventor)

All nature's diff'rence Keeps all nature's peace.

—Alexander Pope

knowledge of geometry can enhance our wonder at the familiar.

Appreciating nature as we ordinarily do is like walking through a city at night amidst its various lights. Each light is beautiful and serves a local purpose. But if we fly in a jet above the city we can look down and see the *patterns* that all the lights make strung along roads and streets to homes and commercial centers, comprehensible patterns we cannot perceive from the ground. We gain a larger view of nature, of life, and of ourself, and we can appreciate the unity of variety. The ancient mathematical philosophers called this integration of all nature's patterns the *harmonia mundi*, the harmony of the world.

In a sense, this book has asked the question "Why are things shaped the ways they are?" We have seen that:

◆ Nature's designs are best suited to their particular purpose. By recognizing and understanding them we can interpret nature's problem-solving strategies in that situation.

◆ Nature's forms are the most practical and functional and most efficient in terms of space, materials, energy, and time. Nature's patterns teach us how to get the most from the least.

◆ All nature's forms undergo continual transformation. Every process strives for balance, harmony, and wholeness. Geometry reveals the process through which harmony comes about.

◆ All the natural events and patterns that manifest themselves, whether we understand them or not, had prior approval of natural laws. Nothing manifests or occurs without the universe's consent.

Seeing how geometry shapes nature, you can understand why ancient artists, architects, and craftspeople of many cultures were impressed by its power and its ability to ennoble human creations. The ancients were aware of nature's geometric language and purposefully employed it in their arts, crafts, architecture, philosophy, myth, natural science, religion, and structures of society from prehistoric

*N*o man finds it difficult to return to nature except the man who has deserted nature.

—Seneca the Younger (4 B.C.–65 A.D., Roman statesman and philosopher)

*T*he union of the mathematician with the poet, fervor with measure, passion with correctness, this surely is the ideal.

—William James (1842–1910, American psychologist and philosopher)

*W*e are star-stuff contemplating the stars.

—Carl Sagan

times through the Renaissance. The world today needs scholars and researchers who give the ancients credit for their intelligence and understanding, to view their art and entire cultures in light of its mathematical symbolism.

BETTER LIVING THROUGH GEOMETRY

In this time of rapid change and transition of the roles of traditional institutions, we have the opportunity to restructure education and teach children differently, to expose them to harmony in all its forms, in nature, music, art, and mathematical beauty. Perhaps children steeped in harmony will become generations of adults who will strive to achieve harmony in the world. And perhaps they will transform our relationship with our environmental matrix to treat the soil, water, air, plants, and creatures differently, cooperatively, in ways born of understanding of the whole, respect for its parts, compassion, and common purpose. Comprehending nature's speech will let us listen to what she is telling us in her own native language, which is also our own. If we can see and understand nature as a harmony in which there is room for diversity and in which we participate, we'll *want* to transform ourselves and our relationships to align with that harmony.

We often act as if inner human nature was unconnected with outer nature, and we judge the outer world by one standard, ourselves by another. Familiarity with the principles of geometry can help reconcile this artificial division. The geometry outside us shows us the principles within ourselves. It's time we, as a global whole, relinquish old models of looking and learning and begin to cooperate. Literacy in nature's script dispels the stereotype of nature as disorganized, unintelligible, and hostile. This book is about reshaping our vision and constructing a new perspective aligned with life-facts. Learning nature's language and reading its message helps abolish the attitude of separateness and encourages us to appreciate diversity. It will lead to nothing less than our own transformation as we find all nature's principles within ourselves.

To learn to resee the world in terms of its patterns

requires a shift within us. But once this shift occurs and we see the familiar world in terms of its shapes and principles, a light turns on and the world brightens, comes into sharper relief. Everything speaks its purpose through its patterns. Even without knowing it we use the same designs found in nature. Look at a microscopic diatom and see a cathedral rose window. Ultimately, the same energy that motivates and guides the natural world does the same for us. All universal designs are found in human body proportions, which we have seen can be repackaged to produce the proportions of a crystal, plant, animal, solar system, and galaxy. It is as if the universe is one single organism, motivated by a single power, developing in many ways to gradually become aware of itself through the awareness of the creatures and forces it produces.

Above the entrance to Plato's Academy in ancient Athens were the words "Let none ignorant of geometry enter here." A knowledge of mathematics was prerequisite for entry to the school's teaching. But let us end this book with a switch and say, "Let none ignorant of geometry *exit* here."

I naudible to our deaf mortal ears
The wide world-rhythms wove their stupendous chant
To which life strives to fit our rhyme-beats here,
Melting our limits in the illimitable,
Tuning the finite to infinity.

—Sri Aurobindo Ghose

Credits